能源与动力工程专业
综合英语教程

主编　李丽君

清华大学出版社
北京交通大学出版社
·北京·

内 容 简 介

本书内容既涉及能源与动力专业相关基础理论知识，也兼顾新技术和新动向。全书分为上、中、下三篇，其中上篇为能源与环境篇，包括第一至第四章，介绍核能、氢能的利用和发展情况、能源转换与储存、能源与环境等问题；中篇为能源与动力工程专业通用理论篇，包括第五至第八章，涵盖工程热力学、工程流体力学、传热学、机械工程这四门基础课的基本概念和理论；下篇为能源与动力工程专业课内容，包括第九至第十五章，主要讲述热导管、自动控制基础、发动机设计、火力发电厂锅炉和汽轮机的基本结构和工作原理、空调用制冷技术和空气调节技术原理及应用、燃料与燃烧等内容。

图书在版编目（CIP）数据

能源与动力工程专业综合英语教程 / 李丽君主编. —北京：北京交通大学出版社：清华大学出版社，2020.3（2025.1 重印）

ISBN 978-7-5121-4123-0

Ⅰ. ① 能…　Ⅱ. ① 李…　Ⅲ. ① 能源–英语–高等学校–教材　② 动力工程–英语–高等学校–教材　Ⅳ. ① TK

中国版本图书馆 CIP 数据核字（2020）第 008920 号

能源与动力工程专业综合英语教程
NENGYUAN YU DONGLI GONGCHENG ZHUANYE ZONGHE YINGYU JIAOCHENG

责任编辑：陈建峰
出版发行：清 华 大 学 出 版 社　　邮编：100084　　电话：010-62776969　　http://www.tup.com.cn
北京交通大学出版社　　邮编：100044　　电话：010-51686414　　http://www.bjtup.com.cn
印 刷 者：北京虎彩文化传播有限公司
经　　销：全国新华书店
开　　本：185 mm×260 mm　　印张：11　　字数：288 千字
版 印 次：2020 年 3 月第 1 版　　2025 年 1 月第 2 次印刷
定　　价：39.00 元

本书如有质量问题，请向北京交通大学出版社质监组反映。对您的意见和批评，我们表示欢迎和感谢。
投诉电话：010-51686043，51686008；传真：010-62225406；E-mail：press@bjtu.edu.cn。

前言

Preface

能源与动力工程致力于传统能源的利用、新能源的开发和更高效地利用能源。其中，能源既包括水、煤、石油等传统能源，也包括核能、风能、生物能等新能源以及未来将广泛应用的氢能。动力工程则涉及内燃机、锅炉、航空发动机、制冷及相关测试技术。本书主体内容既涉及能源动力专业相关基础理论知识，也包括能源工程和动力工程主干专业课程知识及新技术和新动向，目标读者主要为普通高等院校能源与动力工程专业及相近专业本科生及研究生，同时本书也可供能源动力类工程技术人员参考。

为帮助读者在了解专业课程相关知识的同时提升综合语言技能、优化学习效果，本书精选了来自国内外权威学术刊物及专业网站的语篇作为主要语言输入材料，讲述和呈现本领域的基础理论知识和最新发展动态，其来源广泛，内容系统性强，涉及的专业词汇覆盖面广。书中各章均包括 A，B 两篇文章，其中 A 篇可作课堂精读语篇分析讲解，B 篇可以作为课堂泛读材料或者课后补充阅读材料。篇章中所涉及的专业术语均提供解释，方便读者理解文意和扩展专业知识。此外，本书各章均结合本单元主课文设计了相应的科技英语语法精要和练习，帮助使用者切实提高语言基本功和应用能力，真正实现从输入到理解再到产出的语言学习过程。

本书由李丽君担任主编，赵迪、于妍担任副主编。美国印第安纳大学 Raymond Smith 教授在本书的整体章节设置环节提供了帮助，特此感谢。在本书的编写过程中，编者阅读和参考了其他同类教材和网络上的相关资源，这些文献均已在书末参考文献中列及，在此谨向相关文献的作者表示衷心的感谢。由于编者水平有限，加上编写时间仓促，书中难免会有不妥之处，恳请各位同行及广大读者指正。

为方便教师教学，本书配有教学课件及习题答案，可发邮件至 cbswce@jg.bjtu.edu.cn 索取。

编者

2020.1

Section A Energy Resources

Section B General Theories in Energy and Power Engineering

Section C Power Engineering

Section A

Energy Resources

Unit 1

Hydrogen Production Technology

Text A

1. **Hydrogen** is the simplest and most **abundant** element on earth. Hydrogen combines readily with other chemical elements, and it is always found as part of another substance, such as water, hydrocarbon, or alcohol. Hydrogen is also found in natural biomass, which includes plants and animals. For this reason, it is considered as an energy carrier and not as an energy source.

Energy Sources and Hydrogen Production

2. Hydrogen can be produced using diverse, domestic resources, including nuclear, natural gas and coal, biomass, and other renewable sources. The latter include solar, wind, **hydroelectric**, or **geothermal** energy. This diversity of domestic sources makes hydrogen a promising and unending energy source. In order to ensure an inexhaustible supply, it is desirable that hydrogen be produced using a variety of resources and process technologies or pathways. The production of hydrogen can be achieved via various process technologies, including thermal (natural gas reforming, renewable liquid and bio-oil processing, biomass, and coal **gasification**), **electrolytic** (water splitting using a variety of energy resources), and **photolytic** (splitting of water using sunlight through biological and electrochemical materials) technology.

3. The annual production of hydrogen worldwide is estimated to be about 55 million tons with its consumption increasing by approximately 6% per year. Hydrogen can be produced in many ways from a broad **spectrum** of initial raw materials. Nowadays, hydrogen is mainly produced by the steam reforming of natural gas, a process which leads to massive emissions of greenhouse gases. Close to 50% of the global demand for hydrogen is

currently generated via steam reforming of natural gas, about 30% from oil/**naphtha** reforming from refinery/chemical industrial off-gases, 18% from coal gasification, 3.9% from water electrolysis, and 0.1% from other sources. Electrolytic and **plasma** processes demonstrate a high efficiency for hydrogen production, but unfortunately they are considered as energy intensive processes.

4. The fundamental question lies in the development of alternative technologies for hydrogen production to those based on fossil fuels, especially for its utilization as a fuel in the transportation sector. This problem can be faced by the **utilization** of alternative renewable resources and related methods of production, such as the gasification or **pyrolysis** of biomass, electrolytic, photolytic, and thermal cracking of water. However, it is not possible to consider only the ecological perspective, since for example, photolytic cracking of water is environmentally friendly but its efficiency for industrial use is very low. It is thus clear that the processes to be taken into account must consider not only environmental concerns but also the most favorable economics.

Hydrogen from Fossil Fuels

5. Fossil fuel processing technologies **convert** hydrogen-containing materials derived from fossil fuels, such as gasoline, hydrocarbons, **methanol**, or **ethanol**, into a hydrogen-rich gas stream. Fuel processing of **methane** (natural gas) is the most common commercial hydrogen production technology today. Most fossil fuels contain a certain amount of **sulfur**, the removal of which is a significant task in the planning of hydrogen-based economy.

6. Hydrogen gas can be produced from **hydrocarbon fuels** through three basic technologies: (i) steam reforming① (SR), (ii) partial oxidation② (POX), and (iii) autothermal reforming③ (ATR). These technologies produce a great deal of carbon monoxide (CO). Thus, in a subsequent step, one or more chemical reactors are used to largely convert CO into carbon dioxide (CO_2) via the water-gas shift (WGS) and PREFERENTIAL OXIDATION (PROX) or methanation reactions, which are described later.

Economic Aspects of Hydrogen Production

7. At present, the most widely used and cheapest method for hydrogen production is the steam reforming of methane (natural gas). This method includes about half of the world's

① 蒸汽转化，又称水蒸气重整，即天然气在高温、高压下与水蒸气反应制取氢。

② 部分氧化，即天然气催化部分氧化制合成气或氢。

③ 自热重整，即烃类物质在有水蒸气和氧气共同作用的条件下生成合成气的重整过程。

hydrogen production, and cost of production with hydrogen is about 7 USD/GJ[①]. A comparable price for hydrogen is provided by partial oxidation of hydrocarbons. However, greenhouse gases generated by thermochemical processes must be captured and stored, and thus, an increase in the hydrogen price by 25%–30% must be considered.

8. The additionally used thermochemical processes include gasification and pyrolysis of biomass. The price of hydrogen thus obtained is about three times greater than the price of hydrogen obtained by the SR process. Therefore, these processes are generally not considered as cost competitive of steam reforming. The price of hydrogen from gasification of biomass ranges from 10–14 USD/GJ and that from pyrolysis 8.9–15.5 USD/GJ. It depends on the equipment, availability, and cost of **feedstock**.

9. Electrolysis of water is one of the simplest technologies for producing hydrogen without byproducts. Electrolytic processes can be classified as highly effective. On the other hand, the input electricity cost of electrolysis is relatively high and plays a key role in the production costs of hydrogen obtained.

10. By the year 2030, the dominant methods for hydrogen production will be steam reforming of natural gas and **catalyzed** biomass gasification. To a relatively small extent both coal gasification and electrolysis will be used. The use of solar energy in a given context is questionable but also possible. In all likelihood, the role of solar energy will increase by 2050.

(Source: https://www.hindawi.com/archive/2013/690627/)

New Words and Expressions

abundant	*adj.*	丰富的，充裕的
catalyze	*v.*	催化
convert	*v.*	转变，变换
electrolytic	*adj.*	电解的，电解质的
ethanol	*n.*	乙醇
feedstock	*n.*	原料，给料
gasification	*n.*	气化
geothermal	*adj.*	地热的，地温的
hydrocarbon fuel		碳氢燃料，烃类燃料
hydroelectric	*adj.*	水力发电的
hydrogen	*n.*	氢
methane	*n.*	甲烷

① 1 GJ = 10^9 J。

methanol	*n.*	甲醇（methyl alcohol）
naphtha	*n.*	石脑油，粗汽油
photolytic	*adj.*	光解的
plasma	*n.*	等离子体
pyrolysis	*n.*	热解，高温分解
spectrum	*n.*	光谱，波谱
sulfur	*n.*	硫黄
utilization	*n.*	利用，采用

Comprehension Checks

Work in pairs and discuss the following questions. Then share your ideas with the rest of the class.

1. Why is it preferable to view hydrogen as an energy carrier other than an energy source? (para.1)
2. What are some of the process technologies used for producing hydrogen? (para.2)
3. What might be an essential issue in developing hydrogen production technology? And why? (para.4)
4. What might be a major advantage of the steam reforming of methane as a method for hydrogen production? (para.7)
5. What method of hydrogen production is considered highly effective? (para.9)

Vocabulary Tasks

I. Complete the sentences below. The first letter of the missing word has been given.

1. The p________ of hydrogen can be achieved via various process technologies.
2. Close to 50% of the global demand for hydrogen is currently g________ via steam reforming of natural gas.
3. The fundamental question lies in the development of a________ technologies for hydrogen production to those based on fossil fuels.
4. Greenhouse gases generated by thermochemical processes must be c________ and stored.
5. Electrolysis of water is one of the simplest technologies for producing hydrogen without b________.

II. Match each word with the correct definition.

1. substance a. biological material derived from living, or recently living organisms (animals/plants)

2. hydrogen — b. the chemical process by which an atom or group of atoms loses electrons
3. catalyze — c. relating to the most basic and important parts of something
4. availability — d. the difference between the amount of energy that is put into a machine in the form of fuel, effort, etc. and the amount that comes out of it in the form of movement
5. efficiency — e. to make a chemical reaction happen or happen more quickly by acting as a catalyst
6. biomass — f. a chemical element that is the lightest gas, has no color, taste, or smell, and combines with oxygen to form water
7. oxidation — g. the fact that something can be bought, used, or reached, or how much it can be
8. fundamental — h. material with particular physical characteristics

III. Fill in the blanks with the words given below. Change the forms where necessary.

undergo	commercial	byproduct	reform	feedstock
yield	catalyst	oxidation	dominant	presence

There are four main sources for the ____1____ production of hydrogen: natural gas, oil, coal, and electrolysis, which account for 48%, 30%, 18% and 4% of the world's hydrogen production respectively. Fossil fuels are the ____2____ source of industrial hydrogen. Hydrogen can be generated from natural gas with approximately 80% efficiency or from other hydrocarbons to a varying degree of efficiency. Specifically, bulk hydrogen is usually produced by the steam ____3____ of methane or natural gas. The production of hydrogen from natural gas is the cheapest source of hydrogen currently. This process consists of heating the gas in the ____4____ of steam and a nickel ____5____. The resulting exothermic reaction breaks up the methane molecules and forms carbon monoxide CO and hydrogen H_2. The carbon monoxide gas can then be passed with steam over iron oxide or other oxides and ____6____ a water gas shift reaction. This last reaction produces even more H_2. The downside to this process is that its major ____7____ are CO, CO_2 and other greenhouse gases. Depending on the quality of the ____8____ (natural gas, rich gases, naphtha, etc.), one ton of hydrogen produced will also produce 9 to 12 tons of CO_2.

For this process at high temperatures (700–1100 °C), steam (H_2O) reacts with methane (CH_4) in an endothermic reaction to ____9____ syngas.

$$CH_4 + H_2O \rightarrow CO + 3H_2$$

In a second stage, additional hydrogen is generated through the lower-temperature, exothermic, water gas shift reaction, performed at about 360 °C:

$$CO + H_2O \rightarrow CO_2 + H_2$$

Essentially, the oxygen atom (O) is stripped from the additional water (steam) to oxidize CO to CO_2. This ____10____ also provides energy to maintain the reaction. Additional heat required to drive the process is generally supplied by burning some portion of the methane.

Text B

1. Current hydrogen production comes primarily from processing natural gas, although with the **substantial** advances in fuel cells there is increased attention to other fuels such as methanol, propane, gasoline, and logistic fuels such as jet-A, diesel, and JP8 fuels. With the exception of methanol, all of these fuels contain some amount of sulfur, with the specific sulfur species dependent on the fuel type and source.

2. The typical approaches to organo-sulfur removal can be categorized as chemical reaction technologies and adsorptive technologies. Chemical reaction approaches include **hydrodesulfurization** (HDS) and **alkylation**. Most commercial large-scale applications use HDS; therefore, substantial process and catalyst optimization has occurred. In this process, HDS catalysts partially or completely **hydrogenate** the sulfur-bearing molecules, resulting in a release of sulfur as H_2S. The second chemical reaction approach, selective alkylation of organo-sulfur molecules, has been demonstrated at the pilot scale, but to date it has not been implemented on a large commercial scale. This technology increases the molecular weight of the sulfur bearing molecules which increases the boiling point of the compound made up of these molecules. This enables distillation approaches to remove the sulfur. This approach does not require high-pressure hydrogen, which is a potential advantage over HDS. However, the **olefin** content in the fuel will vary, and it may be necessary to intentionally add olefins (or alcohols) to the fuel to convert all the sulfur-bearing molecules and to achieve the desired physical and chemical characteristics of the fuel. There is some evidence that the alkylation process may occur on a limited basis in the course of HDS operation.

3. As the name implies, adsorptive approaches employ **adsorbents** for sulfur removal from the fuel. Most often this is achieved by (1) adsorption of the entire sulfur containing molecule in activated carbon, modified **zeolites** or other materials or (2) adsorption onto metal surfaces such as nickel, wherein nickel **sulfide** is formed, and the remainder of the hydrocarbon is recovered. The former approach is conceptually quite simple to operate, as it can in principle be carried out at **ambient** temperature and pressure using conventional fixed-bed equipment.

4. The other approach is more complicated, requiring a fluid bed operating at elevated

temperatures and pressures. Adsorptive approaches suffer from limited capacity of the material. Adsorbent implementation to hydrocarbon fuels with high levels of sulfur such as JP8 fuels and diesel would require significant quantities of adsorbent, dual beds (to allow simultaneous adsorption and regeneration) with switching between beds, as well as significant logistical issues associated with disposal of spent adsorbents. For low sulfur fuels (<50 ppm sulfur) such as natural gas, adsorbent technologies can make sense depending on the adsorbent capacity and the reactor's capacity. For gas phase sulfur, such as contained in natural gas, activated carbon is the absorbent of choice. Finally, the sorbent materials tend to be very selective to the types of sulfur containing molecules they adsorb.

(Source: J.D. Holladay, et al. 2009. Overview of hydrogen production technologies, Catalysis Today, (139): 224-260)

New Words and Expressions

adsorbent	*n.*	吸附剂
alkylation	*n.*	烷基化
ambient	*adj.*	周围的，外界的
hydrodesulfurization	*n.*	加氢脱硫
hydrogenate	*v.*	使……与氢化合
olefin	*n.*	烯烃
substantial	*adj.*	大量的，实质的
sulfide	*n.*	硫化物
zeolite	*n.*	沸石

Comprehension Questions

1. What is the main reason for the large-scale application of HDS?
   A. It involves substantial process and catalyst optimization.
   B. It releases sulfur as H_2S through hydrogenating the sulfur-bearing molecules.
   C. It enables distillation approaches to remove the sulfur.
2. What does the phrase "at the pilot scale" (*line 6, para.2*) probably mean?
   A. A small study.
   B. A leading research.
   C. A successful test.
3. Why is it necessary to add olefins to the fuel at alkylation?
   A. To acquire high-pressure hydrogen.
   B. To improve physical and chemical features of the fuel.

C. To increase the molecular weight of the sulfur bearing molecules.

4. Which of the following is TRUE about adsorptive approaches?

A. Adsorption of the entire sulfur containing molecule is more feasible technically speaking.

B. Adsorption onto metal surfaces helps to recover the absorbents.

C. Such approaches always rely on a material that is able to take in liquids easily.

5. One condition required for adsorbent implementation with JP8 fuel is ________.

A. a fluid bed

B. activated reactors

C. a large number of adsorbents

Grammar Coach

Derivative

派生词（derivative）是英语语法中的主要构词法之一。派生法（derivation）产生的词汇，即借助前缀或后缀，与词根构成的派生词。熟悉词根及常用前缀和后缀的含义，就可以大致推断出单词的意思。通常派生词的词根决定派生词的基本含义，大部分前缀会改变词根的词义，而多数后缀往往会改变词根的词性。

前缀以否定前缀（negative prefix）如 un-、in-、im-、il-、ir-、non-、dis-、mis-、mal-等为主，可使派生词变成词根的反义词。

例如：

efficiency（*n.* 效率）→inefficiency（*n.* 无效率）

accurate（*adj.* 精准的、正确的）→inaccurate（*adj.* 不准确的）

除了否定前缀，其他常用的前缀还有 anti-、auto-、bi-、co-、counter-、de-、ex-、inter-、mono-、post-、pre-、pro-、re-、sub-、super-、trans-、tri-、ultra-等。

例如：

coexistence（*n.* 共存），deemphasize（*v.* 淡化），superconductive（*adj.* 超导的），ultra-smart（*adj.* 超能的）

科技英语中常见前缀：

aero-（航空的），anti-（反对的），astro-（天体的），bio-（生物的），

calori-（热的），electro-（电的），extra-（超级的），giga-（十亿、千兆），

hydro-（水的），macro-（宏观的、大型的），micro-（微型的、微观的），nano-（极小的），

neuro-（神经），poly-（多的），sym-/syn-（共、合），ultra-（超越的），等等。

相反，加上后缀的词，不但词义有些改变，词性也会不同。动词或形容词加上适

当的后缀之后，可以得到名词派生词。

例如：

equip（*v.* 装备）→equipment（*n.* 设备）

condensed（*adj.* 浓缩的）→condensation（*n.* 浓缩剂）

科技英语中常见后缀如下表所示。

科技英语中常见后缀

后缀	含义	举　　例
-ics	学科，性能	graphics（制图学）
-ism	状态，机制	mechanism（机制）
-ist	学者，倡导者	environmentalist（环境学家）
-ity	状态，性能	abnormality（反常，性能的改变）
-ology	学科，理论	geology（地质学）
-al	……的	optical（光的）
-ed	……的	concatenated（连接的）
-ic	……的	adiabatic（绝热的）
-ate	使……，做……	authenticate（鉴别）
-ify	使……，做……	amplify（扩大，增加）
-ize	使……，做……	miniaturize（小型化）

Exercise

Match the derivatives with their corresponding meanings.

1. parameter	a. the use of machines that operate automatically
2. geology	b. the state in which the reactants and products do not change because the rate of forward reaction is equal to the rate of reverse reaction
3. microorganism	c. the study of the rocks and similar substances that make up the earth's surface
4. metallurgy	d. living thing that on its own is too small to be seen without a microscope
5. transmission	e. the use of an electric current to cause chemical change in a liquid
6. refraction	f. the scientific study of the structures and uses of metals
7. electrolysis	g. a set of facts or a fixed limit that establishes or limits how

	something can or must happen or be done
8. automation	h. the machinery that brings the power produced by the engine to the wheels of a vehicle
9. equilibrium	i. the area of physics connected with the action of heat and other types of energy, and the relationship between them
10. thermodynamics	j. the fact of light or sound being caused to change direction or to separate when it travels through water, glass, etc.

Unit 2

Nuclear Energy

Text A

1. Everything around people is made up of tiny objects called atoms. Most of the mass of each atom is concentrated in the center (which is called the nucleus), and the rest of the mass is in the cloud of electrons surrounding the **nucleus. Protons** and **neutrons** are subatomic particles that comprise the nucleus. Under certain circumstances, the nucleus of a very large atom can split in two. In this process, a certain amount of the large atom's mass is converted to pure energy following Einstein's famous formula $E = mc^2$[①], where m is the small amount of mass and c is the speed of light (a very large number). In the 1930s and 1940s, humans discovered this energy and recognized its potential as a weapon. Technology developed in the Manhattan Project[②] successfully used this energy in a chain reaction to create nuclear bombs. Soon after World War II ended, the newly-found energy source found a home in the propulsion of the nuclear navy, providing **submarines** with engines that could run for over a year without refueling. This technology was quickly transferred to the public sector, where commercial power plants were developed and **deployed** to produce electricity. As of April 2017, 30 countries worldwide are operating 449 nuclear reactors for electricity generation and 60 new nuclear plants are under construction in 15 countries.

Two Fundamental Nuclear Processes

- **Fission**

2. **Fission** is the energetic splitting of large atoms such as **uranium** or **plutonium** into

---

① 爱因斯坦质能方程：E 表示能量，m 表示质量，c 表示光速（常量，$c = 299\ 792.458$ km/s），该方程式主要用来解释核变反应中的质量亏损和计算高能物理中粒子的能量。

② 曼哈顿计划：美国陆军部于 1942 年 6 月开始实施利用核裂变反应来研制原子弹的计划，于 1945 年 7 月 16 日成功进行了世界上第一枚核爆炸试验，并按计划制造出两颗实用的原子弹。

two smaller atoms, called fission products. To split an atom, you have to hit it with a neutron. Several neutrons are also released which can go on to split other nearby atoms, producing a nuclear chain reaction of sustained energy release. This nuclear reaction was the first of the two to be discovered. All commercial nuclear power plants in operation use this reaction to generate heat which they turn into electricity.

- **Fusion**

3. **Fusion** is the combining of two small atoms such as hydrogen or **helium** to produce heavier atoms and energy. These reactions can release more energy than fission without producing as many radioactive byproducts. Fusion reactions occur in the sun, generally using hydrogen as fuel and producing helium as waste. This reaction has not been commercially developed yet and is a serious research interest worldwide, due to its promise of nearly limitless, low-pollution, and **non-proliferative** energy.

Capabilities of Nuclear Power

- **Sustainable**

4. There is quite a bit of talk about nuclear fuel (uranium) running low just like oil. Technically, this is a non-issue, as nuclear waste is recyclable. Economically, it could become a major issue. Today's commercial nuclear reactors burn less than 1% of the fuel that is mined for them and the rest of it or so is thrown away (as depleted uranium and nuclear waste). The US recycling program shut down in the 1970s due to **proliferation** and economic concerns. Today, France and Japan are recycling fuel with great success. New technology exists that can greatly reduce proliferation concerns. Without recycling, the "Red Book" published by the IAEA suggests that there are over 200 years of uranium reserves at current demand. There is also a very large supply of uranium dissolved in seawater at very low concentration. No one has found a cheap-enough way to extract it yet, though people have come close. Nuclear reactors can also run on **thorium** fuel.

- **Ecological**

5. In operation, nuclear power plants emit nothing into the environment except hot water. The classic cooling tower icon of nuclear reactors is just that, a cooling tower. Clean water **vapor** is all that comes out. Very little CO_2 or other climate-changing gases come out of nuclear power generation. The nuclear waste can be handled properly and disposed of geologically without affecting the environment in any way.

6. They're safe too. In March 2013, the former NASA scientist James Hansen published a paper showing that nuclear energy has saved a total of 1.8 million lives in its history worldwide just by displacing air pollution that is a known killer. That includes any deaths nuclear energy has been responsible for from its accidents.

- **Independent**

7. With nuclear power, many countries can approach energy independence. Being "addicted to oil" is a major national and global security concern for various reasons. Using electric or plug-in hybrid electric vehicles (PHEVs) powered by nuclear reactors, we could reduce our oil demands by orders of magnitude. Additionally, many nuclear reactor designs can provide high-quality process heat in addition to electricity, which can in turn be used to **desalinate** water, prepare hydrogen for fuel cells, or to heat neighborhoods, among many other industrial processes.

Deficiencies with Nuclear Power

- **Nuclear Waste**

8. When atoms split to release energy, the smaller atoms that are left behind are often left in excited states, emitting energetic particles that can cause biological damage. Some of the longest lived atoms don't decay to stability for hundreds of thousands of years. This nuclear waste must be controlled and kept out of the environment for at least that long. Designing systems to last that long is a **daunting** task— one that has been a major selling point of anti-nuclear groups.

- **Dramatic Accidents**

9. Three major accidents have occurred in commercial power plants: Chernobyl, Three Mile Island, and Fukushima①. Chernobyl was an uncontrolled steam explosion which released a large amount of radiation into the environment, killing over 50 people, requiring a mass **evacuation** of hundreds of thousands of people, and causing up to 4,000 cancer cases. Three Mile Island was a **partial-core meltdown**, where **coolant** levels dropped below the fuel and allowed some of it to melt. No one was hurt and very little radiation was released, but the plant had to close, causing the operating company and its investors to lose a lot of money. Fukushima was a station **black-out** caused by a huge Tsunami. Four neighboring plants lost cooling and the decay heat melted the cores. Radiation was released and the public was evacuated. These three accidents are very scary and keep many people from being comfortable with nuclear power.

- **Cost**

10. Nuclear power plants are larger and more complicated than other power plants. Many redundant safety systems are built to keep the plant operating safely. This complexity causes the **up-front** cost of a nuclear power plant to be much higher than for a comparable coal plant. Once the plant is built, the fuel costs are much less than fossil fuel costs. In general, the older a nuclear plant gets, the more money its operators make. The large capital

---

① 三大核电站事故：切尔诺贝利核电站事故、三里岛核电站事故、福岛核电站事故。

cost keeps many investors from agreeing to **finance** nuclear power plants.

(Source: https://whatisnuclear.com/articles/nucenergy.html)

New Words and Expressions

black-out	*adj.*	一片漆黑的
coolant	*n.*	冷却剂
daunting	*adj.*	令人气馁的
deploy	*v.*	配置，部署
desalinate	*v.*	使……脱盐
evacuation	*n.*	疏散，撤离
finance	*v.*	供给……经费
fission	*n.*	裂变
fusion	*n.*	聚变
helium	*n.*	氦气
meltdown	*n.*	灾难，熔毁
neutron	*n.*	中子
non-proliferative	*adj.*	防（核）扩散的
nucleus	*n.*	核心，原子核
partial-core	*adj.*	局部核心的
plutonium	*n.*	钚
proliferation	*n.*	增殖，扩散
proton	*n.*	质子
submarine	*n.*	潜水艇
thorium	*n.*	钍
up-front	*adj.*	预先的
uranium	*n.*	铀
vapor	*n.*	蒸气

Comprehension Checks

Work in pairs and discuss the following questions. Then share your ideas with the rest of the class.

1. What were some of the impacts the Manhattan Project has exerted? (para.1)
2. How do fission and fusion differ from each other? (para.2-3)

3. What are the three major capabilities of nuclear power? (para.4-7)
4. Would it be possible to make up for the deficiencies of nuclear energy? (para.10-12)
5. Could you share more background information about the three major nuclear accidents?

Vocabulary Tasks

I. Complete the sentences below. The first letter of the missing word has been given.

1. P________ and neutrons are subatomic particles that comprise the nucleus.
2. This technology was quickly transferred to the public sector, where the commercial power plants were developed and d________ to produce electricity.
3. F________ is the energetic splitting of large atoms such as uranium or plutonium into two smaller atoms, called fission products.
4. Fusion is the combining of two small atoms such as hydrogen or h________ to produce heavier atoms and energy.
5. Some of the longest lived atoms don't d________ to stability for hundreds of thousands of years.

II. Match each word with the correct definition.

1. particle	a. a submersible warship usually armed with torpedoes
2. submarine	b. a tiny piece of anything
3. melt	c. the amount per unit size
4. vapor	d. the process whereby heat changes something from a solid to a liquid
5. density	e. so surprisingly impressive as to stun or overwhelm
6. astounding	f. a visible suspension in the air of particles of some substance
7. daunting	g. the type of alcohol in alcoholic drinks, which can also be used as a fuel for cars
8. ethanol	h. discouraging through fear

III. Fill in the blanks with the words given below. Change the forms where necessary.

subatomic	desalinate	deplete	deploy	decay
daunting	vapor	melt	astounding	evacuate

1. Nanotechnology is the science of materials at the molecular or ________ level.
2. On the down side, the risk would be pretty high, and the impact of switching from one application to another could be ________.

3. Firemen helped ________ the approximately 500 passengers on the trains, taking them out through emergency exits and walking them through the subway tunnel.
4. The application will be ________ on both the instances but will be running only in the new instance.
5. Based on the average size and shape of particles in the ash, any initial charging that occurred would have ________ away many times over.
6. The heat from the impact would have ________ a lot of ice, releasing fresh water.
7. All that comes out of the tailpipe is a little trickle of water and water ________.
8. Energy and the technology to ________ water are both expensive, and this means that desalinating water can be pretty costly.
9. The old, thick ice that moved into this region is now beginning to melt out, which could further ________ the Arctic's remaining store of old, thick ice.
10. The pace of discovery, supported by new instruments and missions and innovative strategies by planet seekers, has been ________.

Text B

1. Nuclear has long been considered a great way to generate the power that lights and heats our homes. It can generate electricity without greenhouse gas emissions. However, after a few terrible disasters in nuclear power plants around the world, people have become more and more aware that, when not treated carefully, nuclear power **poses** rather a significant threat to our way of life. There also have been safety and health concerns involved in storing nuclear (**radioactive**) waste. Luckily, though, in recent years the safety **precautions** of working nuclear power plants have become more and more strict and they are now incredibly safe. However, they still generate tons of very **hazardous** waste material each and every year, material that is difficult to shift.

2. Nuclear waste disposal or radioactive waste management is an important part of nuclear power generation and there are a number of very important and strict guidelines that have to be followed by nuclear power plants and other companies to ensure that all nuclear waste is disposed of safely, carefully and with as little damage as possible to life (whether animal or plant). The amount of radioactive material that is left over from nuclear power plants, however, is luckily very small compared to the waste produced by other methods of generating energy, for instance burning coal or gas, but it can be expensive and it must be done absolutely right.

Dangers of Nuclear Waste Disposal

3. Usually, when nuclear waste is disposed of, it is put into storage containers made of steel that is then placed inside a further **cylinder** made of **concrete**. These protective layers prevent the radiation from getting outside and harming the atmosphere or general surroundings of the nuclear waste. It is a relatively easy and inexpensive method of containing very hazardous materials and actually does not need special transportation or to be stored in a particularly special place, for instance. However, there are a number of dangers that surround nuclear waste disposal.

4. *Long Half Life*: The products of nuclear fission have long half lives, which means that they will continue to be radioactive — and therefore hazardous — for many thousands of years. This means that, if anything were to happen to the waste cylinders in which nuclear waste is stored, this material can be extremely **volatile** and dangerous for many years to come. Since hazardous nuclear waste is often not sent off to special locations to be stored, this means that it is relatively easy to find, and if anyone with ill intent were to look for nuclear waste to serve unpleasant purposes, they may well be able to find some and use it.

5. *Storage*: Another problem with nuclear waste disposal that is still being discussed today is the issue of storage. Many different storage methods have been discussed throughout history, with very few being **implemented** because of the problematic nature of storing such hazardous material that will remain radioactive for thousands of years. Amongst the suggestions that were considered are above-ground storage, **ejection** into space, ocean disposal and disposal into ice sheets.

6. Of these, only one was implemented — ocean disposal was actually used by thirteen different countries and was the method of dumping radioactive waste into the oceans in order to get rid of it. Understandably, this practice is no longer implemented.

7. *Affects on Nature*: One of the biggest concerns that the world has with the disposal of nuclear waste is the affect the hazardous materials could have on animal and plant life. Although most of the time the waste is well **sealed** inside huge drums of steel and concrete, sometimes accidents can happen and leaks can occur. Nuclear waste can have drastically bad effects on life, causing **cancerous** growths, for instance, or causing genetic problems for many generations of animals and plants. Not disposing of nuclear waste properly can therefore have huge environmental impacts that can harm many millions of animals and hundreds of animal species.

Effects of Nuclear Waste Disposal

8. If disposed of properly, nuclear waste disposal need not have any negative effects.

Instead, nuclear waste can lie in its storage place for many thousands of years until it is no longer radioactive and dangerous without being disturbed. However, if the nuclear waste is improperly disposed of or if the disposal methods are compromised, there can be serious consequences and effects of nuclear waste disposal.

9. *Accidents*: Although most of the time a lot of emphasis is placed on the safe disposal of nuclear waste, accidents do occur. Throughout history there have unfortunately been a number of examples of times when radioactive material was not disposed of in the proper ways. This has resulted in a number of disastrous situations, including nuclear waste being spread by dust storms into areas that were **populated** by humans and animals and contaminated of water, whether ponds, rivers or even the sea. These accidents can have disastrous knock-on effects for the animals that reside in or around these areas or that rely on the water of lakes or ponds to survive.

10. Drinking water can become **contaminated**, too, which is absolutely disastrous for locals and residents close to the **epicenter** of the disaster. Even if nuclear waste just seeps into the ground, it can eventually get into **reservoirs** and other water sources and, from there, can reach the homes of people who unwittingly drink high radioactive material. There are examples of these sorts of accidents from all over the world and from all time periods, with severe accidents happening very rarely but having a huge effect on very many people.

11. *Health Effects*: The biggest concern is the negative effects that can have on the human body when exposed to radiation. Long-term effects to radiation can even cause cancer. It is interesting to know that we are exposed to radiations naturally by living our lives that comes from the ground below us.

(Source: http://www.conserve-energy-future.com/dangers-and-effects-of-nuclear-waste-disposal.php)

New Words and Expressions

cancerous	*adj.*	癌的，像癌的
concrete	*n.*	混凝土，凝结物
contaminated	*adj.*	污染的
cylinder	*n.*	圆筒
ejection	*n.*	排出物
epicenter	*n.*	震中
hazardous	*adj.*	危险的，冒险的
implemented	*adj.*	应用的
intent	*n.*	意图
leak	*v.*	泄漏

populate	*v.*	构成……的人口
pose	*v.*	造成
precaution	*n.*	防范，预防措施
radioactive	*adj.*	放射性的
reservoir	*n.*	水库，蓄水池
seal	*v.*	密封
volatile	*adj.*	爆炸性的，反复无常的

Comprehension Questions

1. What can we learn from the first two paragraphs?
   A. Due to more strict safety precautions, the nuclear power plants are safer now.
   B. Nuclear cannot generate electricity without greenhouse gas emissions.
   C. Compared to the waste produced by other energy-generating methods, the amount of radioactive material that is left over from nuclear power plants is very large.
2. In order to dump radioactive waste, many countries have been implementing ________.
   A. ejection into space
   B. ocean disposal
   C. disposal into ice sheets
3. Why does the passage state that it is easy to find nuclear waste?
   A. Because the nuclear waste is not stored in special location.
   B. Because the nuclear waste is radioactive.
   C. Because the nuclear waste has long lives.
4. What does the word "unwittingly" (line 4, para.10) probably mean?
   A. unwillingly
   B. unsteadily
   C. unknowingly
5. According to the passage, which of the following statements is WRONG?
   A. Accidents would never occur if nuclear waste transportation is treated with caution.
   B. Radioactive materials can be easily volatile and cause a number of problems.
   C. If radioactive material was not disposed of in proper ways, accidents would occur.

Grammar Coach

Abbreviation

缩略词（abbreviation）指一个词或短语的简略书写符号，它是语言的一种简化用词现象。以下通过举例介绍缩略词的常见分类和常用缩略规则。

一般来讲，英语缩略词大体上可以分为以下 4 类：

1. 首字母缩略词（acronym），来源于希腊文 akros 和 onyma，意思是取各词首字母重新组成新词，也就是将词组中各词的首字母连成一个词。它不再保持各个首字母自身的读音，而是可以拼读。

例如：

DOB — date of birth 出生日期

AIDS — acquired immune deficiency syndrome 艾滋病

2. 截短词（clipping），也叫简化词，即把一个词或一个词组中的部分字母删掉构成的新词。它是一种重要的英语缩略词的构词方法。

（1）截头（front-clipping）

例如：

telephone — phone 电话

airplane — plane 飞机

（2）去尾（back-clipping）

例如：

advertisement — ad 广告

laboratory — lab 实验室

（3）去掉头尾留中间（front-back clipping）

例如：

influenza — flu 流行性感冒

coca cola — coke 可乐

3. 拼缀词（blend），指将两个或两个以上的单词合成，并在合成时进行缩略的词。

例如：

telex = teleprinter + exchange 电传

modem = moderator + demoderator 调制解调器

Exercise

Please match the following abbreviations with their corresponding meanings.

1. PWR	a. Canadian deuterium uranium
2. BWR	b. International Atomic Energy Agency
3. CANDU	c. pressurized water reactor
4. IAEA	d. International Nuclear Event Scale
5. INES	e. boiling water reactor
6. WANO	f. Nuclear Magnetic Resonance

7. NMR	g. World Association of Nuclear Operators
8. ISI	h. teleprinter exchange
9. chute	i. Institute for Scientific Information
10. telex	j. parachute

Unit 3

Energy Conservation and Storage

Text A

1. Since the discovery of electricity, effective methods have been sought to store energy for use on demand. Over the last century, the energy storage industry has continued to **evolve** and adapt to changing energy requirements and advances in technology.

2. Energy storage systems provide a wide array of technological approaches to managing our power supply in order to create a more **resilient** energy **infrastructure** and bring cost savings to **utilities** and consumers. To help understand the diverse approaches currently being deployed around the world, the energy storage systems can be divided into six main categories.

Solid State Batteries

3. While historically the contributions of a number of scientists and innovators created our understanding of the forces of electricity, Alessandro Volta① is credited with the invention of the first battery in 1800, and along with it the entire field of **electrochemistry**. On its most basic level, a battery is a device consisting of one or more electrochemical cells that convert stored chemical energy into electrical energy. Each cell contains a positive **terminal**, or **anode**, and a negative terminal, or **cathode**. Electrolytes allow **ions** to move between the electrodes and terminals, which allows current to flow out of the battery to perform work.

4. Advances in technology and materials have greatly increased the reliability and output of modern battery systems, and economies of scale have dramatically reduced the associated cost. Continued **innovation** has created new technologies like

① 亚历山德罗·伏特（1745—1827）：意大利物理学家，因在1800年发明伏打电堆而著名。

electrochemical capacitors that can be charged and discharged **simultaneously** and instantly, and provide an almost unlimited operational lifespan.

Flow Batteries①

5. A flow battery is a type of rechargeable battery where **rechargeability** is provided by two chemical components dissolved in liquids contained within the system and most commonly separated by a **membrane**. This technology is **akin** to both a fuel cell and a battery — where liquid energy sources are **tapped** to create electricity and are able to be recharged within the same system.

6. One of the biggest advantages of flow batteries is that they can be almost instantly recharged by replacing the electrolyte liquid, while simultaneously recovering the spent material for **re-energization.**

7. Different classes of flow cells (batteries) have been developed, including **redox, hybrid** and membraneless. The fundamental difference between conventional batteries and flow cells is that energy is stored as the electrode material in conventional batteries but as the electrolyte in flow cells.

Flywheels

8. A flywheel is a rotating mechanical device to store kinetic energy that can be called up instantaneously. At the most basic level, a flywheel contains a spinning mass in its center that is driven by a motor; and when energy is needed, the **spinning** force drives a device similar to a **turbine** to produce electricity, slowing the rate of rotation. A flywheel is recharged by using the motor to increase its rotational speed once again.

9. Flywheel technology has many beneficial properties that enable us to improve our current electric **grid**. A flywheel is able to capture energy from **intermittent** energy sources over time, and deliver a continuous supply of uninterrupted power to the grid. Flywheels also are able to respond to grid signals instantly, delivering frequency regulation and electricity quality improvements.

10. Flywheels were traditionally made of steel and **rotated** on conventional bearings; these were generally limited to a revolution rate of a few thousand RPM②. Modern flywheels are made of carbon fiber materials, stored in **vacuums** to reduce drag, and employ **magnetic** bearings, enabling them to revolve at speeds up to 60,000 RPM.

---

① 流动电池。
② 转速单位，每分钟的转数。

Compressed Air Energy Storage

11. Compressed air energy storage (CAES)[①] is a way to store energy generated at one time for use at another time. At utility scale, energy generated during periods of low energy demand (off-peak) can be released to meet higher demand (peak load) periods.

12. Since the 1870's, CAES systems have been deployed to provide effective, on-demand energy for cities and industries. While many smaller applications exist, the first utility-scale CAES system was put in place in the 1970's with over 290 MW **nameplate** capacity. CAES offers the potential for small-scale, on-site energy storage solutions as well as larger installations that can provide **immense** energy reserves for the grid.

Thermal

13. Take for example modern solar **thermal** power plants, which produce all of their energy when the sun is shining during the day. The excess energy produced during peak sunlight is often stored in these facilities — in the form of molten salt or other materials — and can be used in the evening to generate steam to drive a turbine to produce electricity. Alternatively, a facility can use "off-peak" electricity rates which are lower at night to produce ice, which can be incorporated into a building's cooling system to lower demand for energy during the day.

14. A well designed **thermos** or cooler can store energy effectively throughout the day, in the same way thermal energy storage is an effective resource at capturing and storing energy on a temporary basis to be used at a later time.

Pumped Hydro-Power

15. Gravity is a powerful, inescapable force that surrounds us at all times, and it also **underpins** one of the most established energy storage technologies, pumped hydro-power. Currently the most common type of energy storage is pumped hydroelectric facilities, and we have employed this utility-scale gravity storage technology for the better part of the last century in the United States and around the world.

16. A hydroelectric dam relies on water **cascading** down through a turbine to create electricity to be used on the grid. In order to store energy for use at a later time, there are a number of different projects that use pumps to **elevate** water into a **retained** pool behind a dam, creating an on-demand energy source that can be **unleashed** rapidly. When more energy is needed on the grid, that pool is opened up to run through turbines and produce electricity.

---

① 压缩空气蓄能。

17. But the material that is raised to a higher elevation doesn't have to be water. Companies are currently creating gravitational systems that move **gravel** up the side of a hill and use the same underpinning principle — when energy is needed, the gravel is released and the weight drives a mechanical system that drives a turbine and generates electricity.

(Source: http://energystorage.org)

New Words and Expressions

akin	*adj.*	与……相似的，同类的
anode	*n.*	阳极，正极
cascade	*v.*	倾泻
cathode	*n.*	阴级，负级
electrochemistry	*n.*	电化学
elevate	*v.*	提高，增加
evolve	*v.*	发展，进化
gravel	*n.*	砂砾，碎石
grid	*n.*	网格，输电网
hybrid	*n.*	混合物
infrastructure	*n.*	基础设施，公共建设
immense	*adj.*	巨大的
innovation	*n.*	创新，革新
intermittent	*adj.*	间歇的，断断续续的
ion	*n.*	离子
magnetic	*adj.*	有磁性的
membrane	*n.*	膜，隔膜
nameplate	*n.*	标示牌
rechargeability	*n.*	可再充性
redox	*n.*	氧化还原反应，氧化还原剂
re-energization	*n.*	重新供能
resilient	*adj.*	有弹力的，能复原的
retained	*adj.*	保留的
rotate	*v.*	使旋转
simultaneously	*adv.*	同时地
spin	*v.*	旋转，纺织
tap	*v.*	产生，拔出
terminal	*adj.*	末端的，终点的
thermal	*adj.*	热的

thermos *n.* 热水瓶
turbine *n.* 涡轮
underpin *v.* 巩固，支持
unleash *v.* 释放
utility *n.* 公共事业，实用工具
vacuum *n.* 真空（装置）

Comprehension Checks

Work in pairs and discuss the following questions. Then share your ideas with the rest of the class.

1. How would you comment on the six main categories of the energy system?
2. In what way do flywheels help to improve the current electric grid? (para.8-10)
3. Does a CAES system have a potential to provide large-scale power reserves? (para.11-12)
4. How does a thermos capture heat and cold to create energy on demand? (para.13-14)
5. What is the most common type of energy storage currently? State your reasons. (para.15-17)

Vocabulary Tasks

I. Choose the correct answer from the three choices marked A, B, and C.

1. Electrolytes allow ________ to move between the electrodes and terminals, which allows current to flow out of the battery to perform work.
   A. ions B. iron C. electron
2. When energy is needed, the spinning force drives a device similar to a ________ to produce electricity, slowing the rate of rotation.
   A. tube B. turbine C. flywheel
3. CAES offers the potential for small-scale, on-site energy storage solutions as well as larger installations that can provide ________ energy reserves for the grid.
   A. immortal B. immediate C. immense
4. Flywheel technology has many beneficial properties that enable us to improve our current electric ________.
   A. power B. grid C. generator
5. Modern solar ________ power plants produce all of their energy when the sun is shining during the day.
   A. thermal B. system C. electric

II. Match each word with the correct definition.

1. cascade	a. a creation resulting from study and experimentation
2. elevate	b. to flow or hang in large amounts
3. potent	c. having great influence
4. infrastructure	d. giving a promotion to or assign to a higher position
5. innovation	e. of or relating to or caused by magnetism
6. magnetic	f. an alloy of iron with small amounts of carbon
7. thermos	g. the basic structure or features of a system or organization
8. steel	h. vacuum flask that preserves temperature of hot or cold drinks

III. Fill in the blanks with the words given below. Change the forms where necessary.

akin	underpin	utility	simultaneously	immense
hybrid	grid	spin	intermittent	unleash

1. Energy Star is a platform for retailers, ________, and others to deliver energy-efficient products and services to customers with greater credibility and overall effectiveness.
2. The taste is ________ to walking through a burnt-out building, eating a smoked ham and smoked cheese sandwich.
3. Analysts believe the new data center could be used to ________ new music and video streaming services.
4. Partly because of high gas prices, consumers are taking a second look at ________ cars.
5. We need to ________ massive private investment in clean energy.
6. Separately and ________, a commenting module can associate free from commentary, discussion, and debate with any node.
7. A smart grid would lessen demand overall, allowing for ________ alternative sources of power like wind and solar to be more easily slotted into the electrical supply.
8. Where they are joined to the land, ________ cracks or crevasses can develop due to ocean tides and currents.
9. The higher the wind speed, the faster the turbine ________ and the more electricity is produced.
10. Data ________ provide a single coherent cache shared by all of the application server instances in a cluster.

Text B

1. Energy conservation is not about making limited resources last as long as they can,

which would mean you are doing nothing more than prolonging a crisis. Conservation is the process of reducing demand on a limited supply and enabling that supply to begin to rebuild itself. Many times the best way of doing this is to replace the energy used with an **alternate**.

2. In the case of fossil fuels, the conservation also can include finding new ways to tap into the Earth's supply so that the commonly used oil fields are not drained completely. This allows for those fields to **replenish** themselves more. This is not a process that happens overnight. When you are talking about replenishing natural resources, you are talking about **alleviating** excess demand on the supply in 100 years' time to allow nature to recover.

3. Without energy conservation, the world will **deplete** its natural resources. While some people don't see that as an issue because it will take many decades to happen and they foresee that by the time the natural resource is gone there will be an alternative; the depletion also comes at the cost of creating an enormous **destructive** waste product that then impacts the rest of life. The goal with energy conservation techniques is to reduce demand, protect and replenish supplies, develop and use alternative energy sources, and clean up the damage from the prior energy processes.

4. Below are the energy conservation techniques that can help you reduce your overall carbon footprint and save money in the long run.

5. *Install CFL*[①] *Lights*: Try replacing **incandescent** bulbs in your home with CFL **bulbs**. CFL bulbs cost more **upfront** but last 12 times longer than regular incandescent bulbs. CFL bulbs will not only save energy but over time you end up saving money.

6. *Lower the Room Temperature:* Even a slight decrease in room temperature by only a degree or two, can result in big energy savings. The more the difference between indoor and outdoor temperature, the more energy it consumes to maintain room temperature. A smarter and more comfortable way of doing this is to buy a programmable **thermostat**.

7. *Fix Air Leaks:* Proper **insulation** will fix air leaks that could be costing to you. During winter months, you could be letting out a lot of heat if you do not have a proper insulation. You can fix those leaks yourself or call an energy expert to do it for you.

8. *Use Maximum Daylight:* Turn off lights during the day and use daylight as much as possible. This will reduce the burden on the local power grid and save you a good amount of money in the long run.

9. *Get Energy Audit Done:* Getting energy audit done by hiring an energy audit expert for your home is an energy conservation technique that can help you conserve energy and save good amount of money every month. Home energy audit is nothing but a process that helps you identify areas in your home where it is losing energy and what steps you can take to overcome them. Implement the tips and suggestions given by those energy experts and

---

① CFL 是 compact fluorescent lamp 的缩写，译为紧凑型荧光灯或日光灯，即通常意义上的节能灯（energy-saving light）。

you might see some drop in your monthly electricity bill.

10. *Use Energy Efficient Appliances:* When planning to buy some electrical appliances, prefer to buy one with Energy Star[①] rating. Energy efficient appliances with Energy Star rating consume less energy and save you money. They might cost you more in the beginning but it is much more of an investment for you.

11. *Drive Less, Walk More and Carpool:* Yet another energy conservation technique is to drive less and walk more. This will not only reduce your carbon footprint but will also keep you healthy as walking is a good exercise. If you go to office by car and many of your colleagues stay nearby, try doing **carpooling** with them. This will not only bring down your monthly bill you spend on fuel but will also make you socially more active.

12. *Switch off Appliances when Not in Use:* Electrical appliances like coffee machines, idle printers, and desktop computers keep on using electricity even when not in use. Just switch them off if you don't need them immediately.

13. *Plant Shady Landscaping:* Shady landscaping outside your home will protect it from intense heat during hot and sunny days and chilly winds during the winter season. This will keep your home cool during summer season and will eventually turn to big savings when you calculate the amount of energy saved at the end of the year.

14. *Install Energy Efficient Windows:* Some of the older windows installed at our homes aren't energy efficient. Double **panel** windows and other **vinyl** frames are much better than single panel windows. Choosing correct **blinds** can save on your power bills.

(Source: http://www.conserve-energy-future.com/energy-conservation-techniques.php)

New Words and Expressions

alleviate	*v.*	缓解
alternate	*n.*	替换物
blind	*n.*	百叶窗
bulb	*n.*	灯泡
deplete	*v.*	耗尽，用尽
destructive	*adj.*	毁灭性的，有害的
incandescent	*adj.*	炽热的，发白热光的
panel	*n.*	板，平板
replenish	*v.*	补充，装满
thermostat	*n.*	恒温器，自动调温器

① 能源之星是美国能源部和美国环保署共同推行的一项政府计划，旨在更好地保护生存环境，节约能源。该计划于 1992 年由美国环保署参与，最早在计算机产品上推广。

upfront	*adv.*	在最前面，预先的
vinyl	*n.*	乙烯基

Comprehension Questions

1. What can we learn about energy conservation from the passage?
   A. Energy conservation is to make limited resources last as long as they can.
   B. By the time the natural resource is gone, there must be an alternative.
   C. Without energy conservation, the world will exhaust its natural resources.
2. If you go to work by car and some of your colleagues stay nearby, the author suggests that you can try ________.
   A. walking with them　B. carpooling with them　C. taking a bus with them
3. Which of the following does not qualify an energy-efficient window design?
   A. Vinyl frames.　B. Appropriate blinds.　C. Single panel windows.
4. Which of the following is not mentioned as an advantage of carpooling?
   A. Cut down monthly spending on fuel.
   B. Maintain a healthy living style.
   C. Make people more prosocial.
5. According to the passage, which of the following statements is TRUE?
   A. Transition models stimulate the upgrading of more energy efficient models.
   B. CFL bulbs cost less and last longer than regular incandescent bulbs.
   C. A slight decrease in room temperature cannot make big energy savings.

Grammar Coach

Nominalization

名词化（nominalization）是指把动词、形容词通过一定的方式，如加缀、转化等，转换成名词的语法过程。名词化结构适用于思维精细、逻辑严谨的文章，大量使用名词化结构是科技英语的重要特征之一。

1. 名词化转换形式

（1）动词的名词化：通常行为名词可以由动词加上-ment、-sion、-ion、-ance、-ence等后缀构成。动词的名词化增加了行为名词在科技英语语篇中的出现频率。这类名词除表示行为动作外，还可以表示状态、手段、结果及存在，其中以-ment为后缀的词还能表示事物或工具。

例如：You can <u>rectify</u> this fault if you <u>insert</u> a slash.

→ Rectification of this fault is achieved by insertion of a wedge.

插入楔子可修补这个裂缝。

在这组例子中，名词化发生在两个地方，即 rectify→rectification、insert→insertion，经过名词化处理后，含有两个主谓结构的复合句变成了只含一个主谓结构的简单句，句子结构更加精炼，使用抽象名词替代原来的人称代词作主语使得句子的语体更加正式。

（2）形容词的名词化：作表语的形容词转化为名词。

例如：It is doubtful how accurate the results are.

→ The accuracy of these results is doubtful.

结果的准确性令人怀疑。

（3）从句的名词化：将 if 从句变成作主语的句子。

例如：If we add or remove heat, the state of matter may change.

→ The addition or removal of heat may change the state of matter.

增加或减少热量可以改变物体的状态。

表示行为或状态的名词一般属于抽象名词，科技文体大量使用这类名词，借助抽象思维的逻辑性和概念化，追求表述的简练、凝重、客观和浓缩。

2. 名词化结构的分类

（1）单纯名词化结构。单纯名词化结构指由一个或多个名词修饰一个中心名词构成的名词化结构，属于名词连用的情况，即在中心名词之前用一个或多个名词，它们皆是中心名词的前置修饰成分（在翻译时通常采用顺译法）。

例如：water purification system 净水系统

heat treatment process 热处理过程

computer programming teaching device manual 计算机编程教学仪器指南

（2）复合名词化结构（名词化名词性词组）。复合名词化结构由一个中心名词和形容词、名词、副词、分词及介词短语等多个前置或后置修饰成分构成。修饰成分相对于中心名词的位置是汉英名词性词组在构造上最明显的差别。汉语中修饰成分都是前置式的，英语的中修饰成分有些前置、有些后置。英语中修饰成分的基本语序可概括为：专有性—泛指性—名词，次要意义—重要意义—名词，程度弱—程度强—名词，大—小—名词，即意思越具体、物质性越强，与中心名词的关系越密切的成分就越靠近中心名词。

例如：low average stress values 较低的平均应力值

special strengthening filler material 特殊的强化用的填料

a precise differential air pressure meter 一只精密的差动气压表

（3）由动词派生的名词化结构。这类名词化结构通常是由实义动词派生的名词搭配介词短语构成的，在句中充当主语、宾语或介词短语。一些行为名词与介词后面的宾语有时可构成动宾关系，有时可构成逻辑主谓关系。

例如：Archimedes first discovered the principle of displacement of water by solid bodies.

阿基米德最先发现了固体排水的原理。

句中的名词化结构 displacement of water by solid bodies，由 displace 的名词形式加上两个介词短语构成，用来补充说明 the principle，一方面简化了同位语从句，另一方面强调 displacement 这一事实。

Exercise

Rewrite the following sentences by using nominalization.

1. The paper analyzes the problem and solves it.
2. While it is being compacted, considerable lateral pressure is exerted by the concrete.
3. If the thickness of the lagging is increased, it will reduce the heat losses.
4. The rate of evaporation of the liquid enormously depends on its temperature.
5. We can improve its performance when we use super-heated.

Unit 4

Energy and Environmental Impact

Text A

Environmental Impacts of Renewable Energy Sources

1. Unlike conventional energy sources which require millions of years for formation, renewable energy sources (RES) are sources that constantly renew throughout the human lifespan. Nowadays, RES is a term used for electricity or heat generated from solar, hydro, wind, geothermal, **biomass** or **biogas** energy.

2. The main advantages of RES are **inexhaustibility**, and its impact on the environment. The main disadvantages of RES are low energy density and the technology in development. In fact, the technology of exploitation of RES has become popular only in the last few years, in contrast to conventional power plants that develop from the beginning of electricity production.

Biomass Energy

3. Biomass power plants are based on the same principle as conventional power plants, which differ in the type of fuel. It uses various remains of forestry, agriculture and livestock, which are directly burned or gasified. Unlike other RES, these plants have direct emissions of greenhouse gases and particulate matters, but they take into account continuous renewal of fuel with the amount of produced CO_2 being **equivalent** to consumption in the process of **photosynthesis**. Large land areas are needed to grow biomass in any of the forms which has a direct impact on the eco-world in those areas (**herbicides**, **pesticides**, fertilizers). For the production of **biofuels** (**biodiesel** and **bioethanol**), a large area of **cultivable** land is needed. For the production of biodiesel waste, **edible** oils are also used. Greenhouse gas emissions are

reduced depending on the concentration of biofuels in classic fuels.

Fuel Cells[①]

4. From an environmental point of view, fuel cells are very acceptable because during their work there is no emission of harmful gases and particles. A **byproduct** of fuel cell operation is water produced by a chemical process. They produce electricity and heat simultaneously which is CHP[②]. Such system has operating efficiency up to 90% as waste heat is kept to a minimum.

5. The biggest problem is the production of hydrogen which requires large amounts of electricity produced by fossil fuel power plants. Therefore, negative impacts of fuel cells are often attributed to effects of conventional power plants necessary for the production of hydrogen. A concept has emerged lately about producing hydrogen at night using excess electricity produced by wind turbines (possibly other RES) to completely **eliminate** negative effects.

6. Harmful effects of these plants on humans, animals and the landscape are discussed very often, but all these problems are solved by choosing a good location. The area which is occupied by one unit is 30m^2 and the land between them may still be used for agriculture or **livestock breeding**[③].

Tidal Energy

7. Tidal energy is the use of energy created by raising and lowering sea levels due to the influence of the moon.

8. These plants have no emissions and contribution to **acid** rains. The construction of these power plants, reservoirs and dams can have harmful effects on local **aquatic** life, and on the appearance and structure of the coast. Research conducted in this area shows that the impact on aquatic life is minimal and the impact on local ocean currents is not noticeable. The biggest impact is recorded on the appearance of the site.

Wind Energy

9. The noise produced by the generators is a problem, but there are certain legal norms on the amount of noise they can produce in certain areas, and they are placed only there where they do not exceed the set standards.

10. One of the biggest ecological issues is an impact of wind turbines on bird

① 燃料电池，是一种将燃料与氧化剂的化学能通过电化学反应直接转换成电能的发电装置。

② CHP：combined heat and power（热电联合）。

③ 畜牧业。

populations. Birds can be injured or killed when they fly through **blades** of wind turbines. Current evidence on this issue shows that they offer only a small threat. It could be considered more serious if a **colony** of an endangered species lived in the vicinity of a proposed wind farm, but birds do seem to learn to take account of wind farms.

Wave Energy

11. Wave energy converters remove energy from waves. This means that both near-shore bottom mounted devices and off-shore floating devices will calm the sea. This will be broadly beneficial since it will protect the coast from waves. Wave energy converters will cause some disruption to the marine environment during construction but this should be short lived. Once in place, they should have little impact. There may be a visual impact and **oscillating** water column converters may generate noise from their air turbines. Floating devices are likely to be a hazard to shipping and sites will need to be selected that cause minimum disruption.

Geothermal Plants

12. **Geothermal** plants use fluids drawn from the deep earth. These fluids can carry a mixture of gases, **notably** carbon dioxide (CO_2), hydrogen sulfide (H_2S), methane (CH_4) and **ammonia** (NH_3). These pollutants contribute to global warming, acid rain, and harmful smells if released. Existing geothermal plants emit an average of 122 kg of CO_2 per megawatt-hour (MW • h) of electricity, a small fraction of the emission intensity of conventional fossil fuel plants. Plants that experience high levels of acids and volatile chemicals are usually equipped with emission-control systems to reduce the exhaust. Geothermal plants could theoretically **inject** these gases back into the earth, as a form of carbon capture and storage.

13. Today's power plants that use renewable energy sources are considered more an addition to electricity production. Because of their growing trend, it is very likely that in the next few years they will become the primary source of energy due to their ecological dominance, regional development and the construction of smart grids that are definitely the future of electric power.

(Source: https://bib.irb.hr/datoteka/755710.05-02-14-03.pdf)

New Words and Expressions

acid	*adj.*	酸的
ammonia	*n.*	氨
aquatic	*adj.*	水生的，水栖的

biodiesel	*n.*	生物柴油
bioethanol	*n.*	生物乙醇
biofuel	*n.*	生物燃料
biogas	*n.*	沼气
biomass	*n.*	（单位面积或体积内的）生物量
blade	*n.*	叶片
byproduct	*n.*	副产品
colony	*n.*	群体，集落
cultivable	*adj.*	可耕作的
edible	*adj.*	可食用的
eliminate	*v.*	消除
equivalent	*adj.*	等价的，相等的
herbicide	*n.*	除草剂
inexhaustibility	*n.*	无穷尽
inject	*v.*	注入，增加
notably	*adv.*	显著地，尤其
oscillating	*adj.*	振荡的
pesticide	*n.*	杀虫剂
photosynthesis	*n.*	光合作用

Comprehension Checks

Work in pairs and discuss the following questions. Then share your ideas with the rest of the class.

1. What might be some of the negative impacts of fuel cells? (para. 4-6)
2. How would wind turbines affect bird population? (para. 10)
3. Could you briefly comment on the application of tidal energy? (para. 11)
4. What have you learned about the operation of geothermal plants? (para. 12-13)
5. What do you think of the prospect of developing renewable energy sources in future?

Vocabulary Tasks

I. Choose the correct answer from the three choices marked A, B, and C.

1. The islands are famed for their ________ of sea birds.
   A. families　　B. colonies　　C. groups
2. This single command is ________ to the two previous commands.
   A. equivalent　　B. unequal　　C. comparative

3. Volunteers heard background noises through headphones and had to decide whether ________ sounds of varying frequencies were heard on the right or the left.
   A. stable B. oscillating C. enthusiastic
4. These berries are ________, but those are poisonous.
   A. bitter B. tasty C. edible
5. Traditional handbag makers are ________ more fun into their designs.
   A. injecting B. reducing C. leading
6. God has made devices to keep fish and other ________ animals alive in cold regions.
   A. small B. aquatic C. terrestrial
7. Rhubarb is something of a wonder plant — in addition to an unknown poison in its leaves, they also contain a corrosive ________.
   A. alkali B. material C. acid
8. Wind and photovoltaic power generation are two of the most promising renewable energy technologies owing to many merits they have, such as zero or low emission and ________.
   A. inexhaustibility B. ample C. sufficiency

II. Fill in the blanks with the words given below. Change the forms where necessary.

biofuel	cultivable	geothermal	biogas
notably	eliminate	pesticide	byproduct

1. Greenhouse gas emissions are reduced depending on the concentration of ________ in classic fuels.
2. A concept has emerged lately about producing hydrogen at night using excess electricity produced by wind turbines to completely ________ negative effects.
3. Some regions of the world have done extremely well from globalization, ________ East Asia and China in recent years.
4. The stock of ________ land in industrial countries is more or less fixed, so excess demand resulting from high farm prices and incomes will tend to bid up land rent.
5. ________ plants could theoretically inject these gases back into the earth, as a form of carbon capture and storage.
6. Both natural gas and ________ create emissions when burned, but far less than coal and oil do.
7. A ________ of fuel cell operation is water that is produced by a chemical process.
8. Meat tends to have many other hormones and ________ that can impact your energy levels.

III. Match each word with the correct definition.

1. byproduct — a. a liquid made from vegetable oil or animal fat, which can be used instead of diesel in engines
2. herbicide — b. a substance used to kill unwanted plants
3. biodiesel — c. something that is produced during the manufacture or processing of another product
4. biomass — d. a lake, especially an artificial one, where water is stored before it is supplied to households
5. reservoir — e. the flat wide part of an object that pushes against air or water
6. blade — f. plant and animal matter used to provide power or energy
7. bioethanol — g. a gas that you cannot see or smell, which can be burned to give heat
8. methane — h. a biofuel based on alcohol which may be combined with petrol for use in vehicles

Environmental Impacts of Wind Power

1. **Harnessing** power from the wind is one of the cleanest and most sustainable ways to generate electricity as it produces no toxic pollution or global warming emissions. Wind is also abundant, inexhaustible, and affordable, which makes it available and large-scale alternative to fossil fuels.

2. Despite its vast potential, there are a variety of environmental impacts associated with wind power generation that should be recognized and **mitigated**.

Land Use

3. The land use impact of wind power facilities varies substantially depending on the site: wind turbines placed in flat areas typically use more land than those located in hilly areas. However, wind turbines do not occupy all of this land; they must be spaced approximately 5 to 10 rotor diameters[①] apart (a rotor diameter is the diameter of the wind turbine blades). Thus, the turbines themselves and the surrounding infrastructure (including roads and transmission lines) occupy a small portion of the total area of a wind facility.

4. A survey by the National Renewable Energy Laboratory[②] of large wind facilities in

① 风轮直径，指叶尖旋转圆的直径。

② 国家可再生能源实验室，位于美国科罗拉多州。

the United States found that they use between 30 and 141 acres per megawatt[①] of power output capacity (a typical new utility-scale wind turbine is about 2 megawatts).

5. However, less than 1 acre per megawatt is disturbed permanently and less than 3.5 acres per megawatt are disturbed temporarily during construction. The remainder of the land can be used for a variety of other productive purposes, including livestock **grazing**, agriculture, highways, and hiking trails. Alternatively, wind facilities can be sited on brownfields[②] or other commercial and industrial locations, which significantly reduces concerns about land use.

6. Offshore wind facilities, which are currently not in operation in the United States but may become more common, require larger amounts of space because the turbines and blades are bigger than their land-based counterparts. Depending on their location, such offshore installations may compete with a variety of other ocean activities, such as fishing, recreational activities, sand and gravel extraction, oil and gas extraction, navigation, and **aquaculture**. Employing best practices in planning and siting can help minimize potential land use impacts of offshore and land-based wind projects.

Wildlife and Habitat

7. The impact of wind turbines on wildlife, most notably on birds and bats, has been widely documented and studied. A recent National Wind Coordinating Collaborative (NWCC)[③] review of peer-reviewed research found evidence of bird and bat deaths from collisions with wind turbines and due to changes in air pressure caused by the spinning turbines, as well as from habitat disruption. The NWCC concluded that these impacts were relatively low and did not pose a threat to species populations.

8. Additionally, research into wildlife behavior and advances in wind turbine technology have helped to reduce bird and bat deaths. For example, wildlife biologists have found that bats are most active when wind speeds are low. Using this information, the Bats and Wind Energy Cooperative concluded that keeping wind turbines motionless during times of low wind speeds could reduce bat deaths by more than half without significantly affecting power production. Other wildlife impacts can be mitigated through better siting of wind turbines. The U.S. Fish and Wildlife Services[④] has played a leadership role in this effort by convening an advisory group including representatives from industry, state and tribal governments, and nonprofit organizations that made comprehensive recommendations on appropriate wind

---

① 兆瓦，是一种表示功率的单位，常用来指发电机组在额定情况下单位时间内能发出来的电量。

② 棕色地带是指被废弃的、闲置的或未得到充分利用的工业和商业设施，由于这些设施已存在严重的或潜在的环境污染，因而难以利用或再开发。

③ 国家风能协调委员会。

④ 美国鱼类及野生动植物管理局。

farm siting and best management practices.

9. Offshore wind turbines can have similar impacts on **marine** birds, but as with onshore wind turbines, the bird deaths associated with offshore wind are minimal. Wind farms located offshore will also impact fish and other marine wildlife. Some studies suggest that turbines may actually increase fish populations by acting as artificial **reefs**. The impact will vary from site to site, and therefore proper research and monitoring systems are needed for each offshore wind facility.

Public Health and Community

10. Sound and visual impact are the two main public health and community concerns associated with operating wind turbines. Most of the sound generated by wind turbines is **aerodynamic**, caused by the movement of turbine blades through the air. There is also mechanical sound generated by the turbine itself. Overall sound levels depend on turbine design and wind speed.

11. Some people living close to wind facilities have complained about sound and **vibration** issues, but industry and government-sponsored studies in Canada and Australia have found that these issues do not adversely impact public health. However, it is important for wind turbine developers to take these community concerns seriously by following "good neighbor" best practices for siting turbines and initiating open dialogue with affected community members. Additionally, technological advances, such as minimizing blade surface **imperfections** and using sound-absorbent materials can reduce wind turbine noise.

12. Under certain lighting conditions, wind turbines can create an effect known as shadow flicker. This annoyance can be minimized with careful siting, planting trees or installing window **awnings**, or **curtailing** wind turbine operations when certain lighting conditions exist.

13. The Federal Aviation Administration (FAA) requires that large wind turbines, like all structures over 200 feet high, have white or red lights for aviation safety. However, the FAA recently determined that as long as there are no gaps in lighting greater than a half-mile, it is not necessary to light each tower in a multi-turbine wind project. Daytime lighting is unnecessary as long as the turbines are painted white.

14. When it comes to **aesthetics**, wind turbines can **elicit** strong reactions. To some people, they are graceful **sculptures**; to others, they are **eyesores** that **compromise** the natural landscape. Whether a community is willing to accept an altered skyline in return for cleaner power should be decided in an open public dialogue.

Life-Cycle Global Warming Emissions

15. While there are no global warming emissions associated with operating wind turbines, there are emissions associated with other stages of a wind turbine's life-cycle,

including materials production, materials transportation, on-site construction and assembly, operation and maintenance, and **decommissioning** and **dismantlement**.

16. Estimates of total global warming emissions depend on a number of factors, including wind speed, percent of time the wind is blowing, and the material composition of the wind turbine. Most estimates of wind turbine life-cycle global warming emissions are between 0.02 and 0.04 pounds of carbon dioxide equivalent per **kilowatt-hour**. To put this into context, estimates of life-cycle global warming emissions for natural gas generated electricity are between 0.6 and 2 pounds of carbon dioxide equivalent per kilowatt-hour and estimates for coal-generated electricity are 1.4 and 3.6 pounds of carbon dioxide equivalent per kilowatt-hour.

(Source: http://www.ucsusa.org/clean-energy/renewable-energy/environmental-impacts-wind-power#.WRXOCxhY5-U)

New Words and Expressions

aerodynamic	*adj.*	空气动力学的
aesthetics	*n.*	美学
aquaculture	*n.*	水产养殖，水产业
awning	*n.*	雨篷，天幕
compromise	*v.*	危害
curtail	*v.*	缩减
decommission	*v.*	停止使用
dismantlement	*n.*	拆除
elicit	*v.*	引起
eyesore	*n.*	眼中钉
grazing	*n.*	放牧
harness	*v.*	利用
imperfection	*n.*	缺点，瑕疵
kilowatt-hour	*n.*	千瓦时
marine	*adj.*	海生的
mitigate	*v.*	使缓和，使减轻
reef	*n.*	礁
vibration	*n.*	振动

Comprehension Questions

1. What can we learn about land impact of wind power facilities?

A. Wind turbines are all located in hilly land areas.

B. Wind facilities sited on brownfields can help reduce concerns about land use.

C. Offshore wind facilities don't need vast space.

2. According to the passage, which of the following statements is TRUE about wind turbine on wildlife?

A. The habitat disruption poses a threat to wildlife species populations.

B. Keeping wind turbines motionless during times of low wind speeds could reduce bat deaths.

C. Compared with onshore wind turbines, offshore wind turbines pose a higher risk to wild birds.

3. What does the underlined word "adversely" (line 3, para. 11) probably mean?

A. 不利地　　B. 有效地　　C. 直接地

4. According to FAA, it is unnecessary to light each tower in a multi-turbine wind project because ________.

A. the wind turbines can elicit strong reactions

B. daytime lighting is a kind of waste with the turbines painted light

C. the gaps between lighting are no more than half a mile

5. Estimates of total global warming emissions could be made on the basis of ________.

A. wind direction

B. average temperature difference in a year

C. material composition of the wind turbine

Describing Numbers and Units

1. 数词的表达方式

（1）基数词表示法

① 十位数。表示十位数时，在几十和个位基数词之间加连字符"-"。

例如：21—twenty-one　76—seventy-six

② 百位数。表示百位数时，在几十几与百位基数词之间加 and。

例如：101—one hundred and one　320—three hundred and twenty

③ 千位数及以上。从数字的右端向左端数起，每三位数字前加一个逗号","。从右向左，第一个","前的数字后添加 thousand（千），第二个","前的数字后添加 million（百万），第三个","前的数字后添加 billion（十亿），两个逗号之间最大的数为百位数形式。

例如：2,648 — two thousand six hundred and forty-eight

16,250,064 — sixteen million two hundred and fifty thousand sixty-four

④ 基数词在表示确切的数字时，不可使用百、千、百万、十亿的复数形式；但是当基数词表示不确切数字，如成百、成千上万、三三两两时，则以复数形式出现。

例如：

There are hundreds of people in the hall. 大厅里有数以百计的人。

They went to the theatre in twos and threes. 他们三三两两地来到了剧院。

⑤ 表示人的不确切岁数或年代时，用“几十”的复数形式。

例如：

He became a professor in his thirties. 他三十多岁时就成了教授。

It was in the 1960s. 那是在二十世纪六十年代。

（2）序数词表示法

① 第一至第十九。其中，one—first、two—second、three—third、five—fifth、eight—eighth、nine—ninth、twelve—twelfth 为特殊形式，其他的序数词都是由其相对应的基数词后面加“th”构成。

例如：six—sixth　nineteen—nineteenth

② 第二十至第九十九。

整数第几十的形式由其对应的基数词改变结尾字母 y 为 i 加“eth”构成。

例如：twenty—twentieth　thirty—thirtieth

表示第几十几时，用几十的基数词形式加上连字符“-”和个位序数词形式一起表示。

例如：thirty-first 第三十一　ninety-ninth 第九十九

③ 第一百以上的多位序数词由基数词的形式变结尾部分为序数词形式来表示。

例如：one hundred and twenty-first 第一百二十一

one thousand, three hundred and twentieth 第一千三百二十

（3）分数表示法

① 基数词作分子，序数词作分母，除分子是“1”的情况外，其他情况下序数词都要用复数形式。

例如：

3/4—three fourths 或 three quarters　1/3—one third 或 a third　1/2—a half

1/4—one quarter 或 a quarter　24/25—twenty-four twenty-fifths

② 当分数后面接名词时，如果分数表示的值大于 1，名词用复数；如果分数表示的值小于 1，名词用单数。

例如：$1\frac{1}{2}$ hours、$2\frac{3}{4}$ meters、4/5 meter、5/6 inch

③ 表示“*n* 次方”时，指数用序数词，底数用基数词。

例如：

10 的 7 次方 the seventh power of ten（ten to the seventh power）

6 的 10 次方 the tenth power of six（six to the tenth power）

（4）小数表示法

① 小数用基数词来表示，以小数点为界，小数点左首的数字为一个单位，表示整数，数字合起来读；小数点右首的数字为一个单位，表示小数，数字分开来读；小数点读作 point，0 读作 zero 或 o，整数部分为零时，可以省略不读。

例如：

0.4—zero point four 或 point four　10.23—ten point two three

② 数字值大于 1 时，小数后面的名词用复数；数字值小于 1 时，小数后面的名词用单数。

例如：1.03 meters、0.49 ton

（5）百分数表示法

用基数词+percent 表示。

例如：50%—fifty percent　3%—three percent　0.12%—zero point one two percent

percent 一词中 per 表示“每一”，cent 表示“百”，因此 percent 一词在此不用复数形式。

2. 单位的表达方式

（1）表示长、宽、高、面积等，用基数词+单位词+形容词表示，或者用基数词+单位词+in+名词表示。

例如：

two meters long 或 two meters in length

three feet high 或 three feet in height

four inches wide 或 four inches in width

（2）表示温度时，用 below zero 表示零度以下，温度用基数词+degree(s)+单位词（centigrade 或 Fahrenheit）表示。

例如：

thirty-six degrees centigrade

four degrees below zero centigrade

thirty-two degrees Fahrenheit.

（3）常见的其他英文单位。

bottle 瓶	dozen 打	lot 批	roll 卷
set 套、台	drum 桶	pound 英镑	inch 英寸
foot 英尺	Joule（J）焦耳	volt（V）伏	ampere（A）安
ohm 欧	kilowatt（kW）千瓦	ton 吨	kilogram（kg）千克
sheet 张	box 箱	liter 升	piece 件

Exercise

Describing the following Numbers or Units in English.

1. 2,034
2. 3,123,400
3. 1,050,000,000
4. 第五十
5. 第一千零一
6. 7/8
7. 三分之二英寸
8. 5 英尺高

Section B

General Theories in Energy and Power Engineering

Unit 5

Thermodynamics System

Text A

1. The branch of science called thermodynamics deals with systems that are able to transfer thermal energy into at least one other form of energy (mechanical, electrical, etc.) or into work. The laws of thermodynamics were developed over the years as some of the most fundamental rules which are followed when a thermodynamic system goes through some sort of energy change.

Macroscopic and Microscopic Approach of Thermodynamics

2. Macroscopic Thermodynamics: When matter is considered as continuous function of space variable, it is classical approach of thermodynamics, which requires simple mathematical **formula** for analyzing the system.

3. Microscopic Thermodynamics: All the atoms and **molecules** of the system are considered and the summation of all the atoms and molecules are used. It is statistical approach of thermodynamics.

Thermodynamic System, Surrounding & Boundary

4. A region in space in which investigation is going on or small part of the universe to which we can apply the laws of thermodynamics called as thermodynamic system.

Types of Thermodynamic System

5. Thermodynamic system can be classified as: closed systems, open systems, isolated systems.

6. Closed system are able to exchange energy (heat and work) but not matter with their environment. A greenhouse is an example of a closed system exchanging heat but not work

with its environment. Whether a system exchanges heat, work or both is usually thought of as a property of its boundary.

7. Open systems may exchange any form of energy as well as matter with their environment. A boundary allowing matter exchange is called **permeable**. The ocean would be an example of an open system.

8. Isolated systems are completely isolated from their environment. They do not exchange heat, work or matter with their environment. The only truly isolated system there could be is the universe, but even that is up for debate if the Big Bang① is considered.

Boundary

9. A system boundary is a real or imaginary two-dimensional closed surface that encloses or **demarcates** the volume or region that a thermodynamic system occupies, across which quantities such as heat, mass, or work can flow. In short, a thermodynamic boundary is a geometrical division between a system and its surroundings. **Topologically**, it is usually considered to be nearly or piecewise smoothly **homeomorphic** with a **two-sphere**, because a system is usually considered to be simply connected.

10. Boundaries can also be fixed (e.g. a constant volume reactor) or moveable (e.g. a piston). For example, in a **reciprocating** engine, a fixed boundary means the piston is locked at its position; as such, a constant volume process occurs. In that same engine, a moveable boundary allows the **piston** to move in and out. Boundaries may be real or imaginary. For closed systems, boundaries are real while for open systems, boundaries are often imaginary. For theoretical purposes, a boundary may be declared to be **adiabatic**, **isothermal**, **diathermal**, **insulating**, permeable, or semipermeable, but actual physical materials that provide such idealized properties are not always readily available.

Surroundings

11. The system is the part of the universe being studied, while the surroundings are the remainder of the universe that lies outside the boundaries of the system. They are also known as the environment, and the reservoir. Depending on the type of system, they may interact with the system by exchanging mass, energy (including heat and work), **momentum**, electric charge, or other conserved properties. The environment is ignored in analysis of the system, except in regards to these interactions.

Systems in Equilibrium

12. At thermodynamic equilibrium, a system's properties are, by definition, unchanging

① 某些科学家提出的关于宇宙起源的创世大爆炸。

in time. Systems in equilibrium are much simpler and easier to understand than systems which are not in equilibrium. Often, when analyzing a thermodynamic process, it can be assumed that each intermediate state in the process is at equilibrium. This will also considerably simplify the analysis.

13. In isolated systems it is consistently observed that as time goes on internal rearrangements diminish and stable conditions are approached. Pressures and temperatures tend to equalize, and matter arranges itself into one or a few relatively homogeneous phases. A system in which all processes of change have gone practically to completion is considered to be in a state of thermodynamic equilibrium. The thermodynamic properties of a system in equilibrium are unchanging in time. Equilibrium system states are much easier to describe in a deterministic manner than non-equilibrium system states.

14. In thermodynamic processes, large departures from equilibrium during intermediate steps are associated with increases in **entropy** and increases in the production of heat rather than useful work. It can be shown that for a process to be **reversible**, each step in the process must be reversible. For a step in a process to be reversible, the system must be in equilibrium throughout the step. That ideal cannot be accomplished in practice because no step can be taken without **perturbing** the system from equilibrium, but the ideal can be approached by making changes slowly.

（Source: https://www.slideshare.net/engineeringmasters/thermodynamic-system-43962313）

New Words and Expressions

adiabatic	*adj.*	绝热的
demarcate	*v.*	区别
diathermal	*adj.*	透热辐射的
entropy	*n.*	熵（热力学中表征物质状态的参量之一）
equilibrium	*n.*	力平衡，均衡
formula	*n.*	公式
homeomorphic	*adj.*	同胚的
insulating	*adj.*	绝缘的
isothermal	*adj.*	等温的
molecule	*n.*	分子
momentum	*n.*	动量
permeable	*adj.*	能透过的
perturb	*v.*	扰乱
piston	*n.*	活塞

reciprocating *adj.* 往复的，摆动的
reversible *adj.* 可逆的
topologically *adv.* 拓扑地
two-sphere *n.* 二维球

Comprehension Checks

Work in pairs and discuss the following questions. Then share your ideas with the rest of the class.

1. What are the major differences between macroscopic and microscopic approach of thermodynamics? (para.2-3)
2. How many types of thermodynamic system are introduced in the text? Please give an example for each type. (para.5-8)
3. How is a system boundary defined and what are some of the main features? (para.9-10)
4. What are some possible advantages of equilibrium system? (para.12-14)
5. Could you briefly describe thermodynamic processes? (para.12-14)

Vocabulary Tasks

I. Complete the sentences below. The first letter of the missing word has been given.

1. A________ means that there's no heat transferred between the system and the surroundings.
2. A space suit is lined with special i________ materials that stop you losing the heat inside (produced by your body) to the outside.
3. Venus, earth, and other planets also p________ mercury's motion.
4. R________ compressors can produce a lot of heat.
5. The e________ for the global currency market could be rotating devaluation, which would anchor inflation in all major economies.

II. Choose the correct answer from the three choices marked A, B, and C.

1. As the ________ moves up it is compressed, the spark plug ignites combustion, and exhaust exits through another hole in the cylinder.
   A. piston B. bottle C. pile
2. The lack of ________ surfaces are loading drainage systems and increasing the risk of flooding.
   A. clear B. solid C. permeable
3. The discovery could open a new front in the battle against the bulge if the ________ has

the same effect in humans.

A. particle B. atom C. molecule

4. An ________ process is a change of a system, in which the temperature remains constant: $\Delta T = 0$.

A. isothermal B. accurate C. insulating

5. Topologically, it is usually considered to be nearly or piecewise smoothly ________ with a two-sphere, because a system is usually considered to be simply connected.

A. homogeneous B. homeomorphic C. homeostatic

6. The wheel was allowed to roll down the slope, gathering ________ as it went.

A. material B. momentum C. moment

7. It says that if in this space any closed-curve object, such as a rubber band, can be shrunk to a point, then the space is ________ equivalent to a sphere.

A. topologically B. logically C. absolutely

8. In this paper, basing upon the Plank's law, the calculation ________ about the flame radiant intensity has been deduced directly.

A. formula B. form C. formation

Text B

Laws of Thermodynamics

The Zeroeth Law of Thermodynamics

1. The zeroeth law of thermodynamics (the zeroeth law for short) is sort of a transitive property of thermal equilibrium which claims that two systems in thermal equilibrium with a third system are in thermal equilibrium to each other. The transitive property of mathematics says that if A = B and B = C, then A = C. The same is true of thermodynamic systems that are in thermal equilibrium.

2. One consequence of the zeroeth law is the idea that measuring temperature has any meaning whatsoever. In order to measure a temperature, thermal equilibrium must be reached between the **thermometer** as a whole, the mercury inside the thermometer, and the substance being measured. This, in turn, results in being able to accurately tell what the temperature of the substance is.

3. This law was understood without being **explicitly** stated through much of the history of thermodynamics study, and it was only realized that it was a law in its own right at the

beginning of the 20th century. It was British physicist Ralph H. Fowler[①] who first coined the term "zeroeth law", based on a belief that it was more fundamental even than the other laws.

The First Law of Thermodynamics

4. The first law of thermodynamics (the first law for short) states the change in a system's internal energy is equal to the difference between heat added to the system from its surroundings and work done by the system on its surroundings.

5. If you add heat to a system, there are only two things that can be done — change the internal energy of the system or cause the system to do work (or, of course, some combination of the two). All of the heat energy must go into doing these things.

Mathematical Representation of the First Law of Themodynamics

6. Physicists typically use uniform conventions for representing the quantities in the first law of thermodynamics.

U_1 (or U_i): initial internal energy at the start of the process

U_2 (or U_f): final internal energy at the end of the process

$$\Delta U = U_2 - U_1$$

ΔU: change in internal energy (used in cases where the specifics of beginning and ending internal energies are irrelevant)

Q: heat transferred into ($Q > 0$) or out of ($Q < 0$) the system

W: work performed by the system ($W > 0$) or on the system ($W < 0$)

7. This **yields** a mathematical representation of the first law which proves very useful and can be rewritten in a couple of useful ways:

$$U_2 - U_1 = \Delta U = Q - W$$

$$Q = \Delta U + W$$

8. The analysis of a thermodynamic process, at least within a physics classroom situation, generally involves analyzing a situation where one of these quantities is either 0 or at least controllable in a reasonable manner. For example, in an adiabatic process, the heat transfer (Q) is equal to 0 while in an **isochoric** process the work (W) is equal to 0.

The Second Law of Thermodynamics

9. According to the second law of thermodynamics, it is impossible for a process to have as its sole result the transfer of heat from a cooler body to a hotter one.

10. The second law of thermodynamics is formulated in many ways, as will be addressed shortly, but is basically a law which — unlike most other laws in physics — deals not with

① 拉尔夫·霍华德·福勒（1889—1944）：英国物理学家、天文学家。

how to do something, but rather deals entirely with placing a restriction on what can be done.

11. It is a law that says nature **constrains** us from getting certain kinds of outcomes without putting a lot of work into it, and as such is also closely tied to the concept of the conservation of energy, much as the first law of thermodynamics is.

12. In practical applications, this law means that any heat engine or similar device based upon the principles of thermodynamics cannot, even in theory, be 100% efficient.

13. This principle was first illuminated by the French physicist and engineer Sadi Carnot①, as he developed his Carnot cycle engine in 1824, and was later formalized as a law of thermodynamics by German physicist Rudolf Clausius②.

Entropy and the Second Law of Thermodynamics

14. The second law of thermodynamics is perhaps the most popular outside of the **realm** of physics, because it is closely related to the concept of entropy, or the disorder created during a thermodynamic process. Reformulated as a statement regarding entropy, the second law of thermodynamics reads:

In any closed system, the entropy of the system will either remain constant or increase.

15. In other words, each time a system goes through a thermodynamic process, the system can never completely return to precisely the same state it was in before. This is one definition used for the arrow of time, since entropy of the universe will always increase over time according to the second law of thermodynamics.

The Third Law of Thermodynamics

16. The third law of thermodynamics is essentially a statement about the ability to create an absolute temperature scale, for which absolute zero is the point at which the internal energy of a solid is precisely zero.

17. Various sources show the following three potential formulations of the third law of thermodynamics:

(1) It is impossible to reduce any system to absolute zero in a **finite** series of operations.

(2) The entropy of a perfect crystal of an element in its most stable form tends to be zero as the temperature approaches absolute zero.

(3) As temperature approaches absolute zero, the entropy of a system approaches a constant.

(Source: https://www.thoughtco.com/laws-of-thermodynamics-p3-2699420)

---

① 萨迪·卡诺（1796—1832）：法国青年工程师，热力学的创始人之一。

② 鲁道夫·克劳修斯（1822—1888）：德国物理学家和数学家，热力学的主要奠基人之一。他重新陈述了萨迪·卡诺的定律（又被称为卡诺循环），把热理论研究推至一个更真实、更健全的层次。

New Words and Expressions

constrain *v.* 约束
explicitly *adv.* 明确地
finite *adj.* 有限的，限定的
isochoric *adj.* 等体积的，等容的
realm *n.* 领域，范围
thermometer *n.* 温度计，体温计
yield *v.* 产生，导致

Comprehension Questions

1. Which law was granted as a law in its own right at the beginning of the 20th century?
   A. The zeroeth law of thermodynamics.
   B. The second law of thermodynamics.
   C. The third law of thermodynamics.
2. Which of the following mathematical representation is true with the first law of themodynamics ?
   A. In an adiabatic process, $Q = W = 0$.
   B. $Q + W = \Delta U$
   C. $U_2 - U_1 = Q - W$
3. Which of the following statement is TRUE about the second law of thermodynamics?
   A. In a closed system, the entropy of the system will remain constant or increase.
   B. Although a system goes through a thermodynamic process, it can still return to the same state as before.
   C. In practical applications, any heat engine or similar device based on the principles of the second law of thermodynamics cannot be 100% efficient.
4. What does the underlined word "illuminate" (line 1, para.13) probably mean?
   A. 质疑　　B. 阐述　　C. 反对
5. What can be inferred from the third law of thermodynamics?
   A. The system could be reduced to absolute zero after a series of operations.
   B. When a system reaches absolute zero, molecules stop all movement.
   C. The entropy of a system would be constantly changing when its temperature approaches absolute zero.

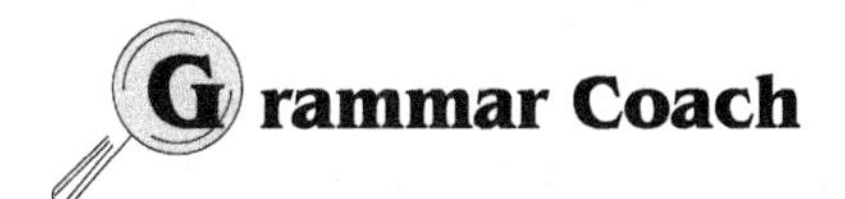

Grammar Coach

Passive Voice

被动语态（passive voice）在科技英语中被广泛使用，因为被动语态能够把所要论证、说明的主要问题放在句子的主语位置上，更加引人关注。此外，被动语态相比主动语态缺少主观色彩，更能凸显事实的客观性，这正是科技英语文体所需要的。因此，在科技英语中，凡是在不需要或无法指出行为主体的场合，或者在需要突出行为客体的场合通常都选择被动语态来表达。

1. 科技英语中被动语态的运用

（1）科技文体强调客观性和理论性，通常很少采用第一人称和第二人称等带有强烈个人情感色彩的词语，如“I think” “we believe”等。

例如：Much of the recent interest in practical quantum computing has been stimulated by the discovery of a quantum algorithm that allows the determination of the prime factors of large composite numbers efficiently.

本句中具体事物作主语，而非第一、二人称，强调客观事实，避免作者在行文中的主观渲染，符合科技文体的特征。

（2）通过被动语态把重要的成分置于句首，可起到强调的作用。与中文的表达习惯相反，英文的语言习惯是在主要内容全部表达清楚后再表达其他成分，被动语态可以使重要信息前置，使得行文简洁流畅。

例如：The role of isotopes as phonon scatters with a resulting influence on the thermal conductivity k was first considered theoretically by Pomeranchuk in 1942.

本句强调的是同位素掺杂声子散射对导热系数 k 产生的影响，利用被动语态将其放置在句子前端，使得读者能够快速、准确地获取重要信息，符合科技文体明确、清晰的特征。

（3）被动语态句式结构明晰，使得文章总体结构更加连贯、紧凑。科技文体的特征在于其行文简洁明了、表述客观事实准确、信息量大、重在强调客观事实。因此，写作或者翻译科技文本类型的文章时，应尽量切合这一文体要求。

例如：This principle was first illuminated by the French physicist and engineer Sadi Carnot, as he developed his Carnot cycle engine in 1824, and was later formalized as a law of thermodynamics by German physicist Rudolf Clausius.

本句由连词 and 连接的两个被动句构成，如果将第二个被动句改为主动句，则行文繁杂，内容突兀。此外，句子重点强调的是 principle 的发展过程，而非施动者。被动语态能够巧妙地解决这一问题，使得文体自然流畅。

2. 科技英语中被动语态的翻译

（1）科技英语中的被动句译成汉语的主动句。当被动句中的主语为无生命的名词且句中没有介词引导的行为主体时，应依据汉语语法习惯及上下文语气，将其译成主动句，原文中的主语仍对应译成译文中的主语，不必使用“……被……”的结构，可以通过在译文的谓语动词前加上“会”“能”“应”等使其符合汉语表达习惯。

例如：The experiment will be finished in two weeks.

不当译文：这项实验将在两周后被完成。

正确译文：这项实验将在两周后完成。

（2）在科技英语中，如果被动句的主语是表示事物的名词，并且句中带有地点状语或由介词 by 短语构成的状语，可将原句中的状语译为汉语中的主语。

例如：They are taught to operate this machine by the engineers.

不当译文：他们被工程师教如何操作这台机器。

正确译文：工程师教他们如何操作这台机器。

（3）在“it is (has been)+过去分词+ that...”句型中，it 为形式主语，代替主语从句，该句型可译成汉语的无主句或不定人称句。译文也可以在动词前增译“人们”“有人”等泛指主语的词。

例如：It is believed that oil is derived from marine plant and animal life.

不当译文：石油来源于海生动植物。

正确译文：人们认为石油来源于海生动植物。

（4）当英文强调句子的被动意义或受动者的动作时，翻译时可以不改变被动语态的使用和句子的原有形式，通过使用其他可以表示被动含义的词语来解释“被”这种状态，如译为“由”“所”“让”“遭到”等。

例如：Copper is extracted from the ore of the blast furnace.

不当译文：铜被鼓风炉从矿石中提炼出来。

正确译文：铜是用鼓风炉从矿石中提炼出来的。

（5）当无法确定动作的发出者时，可以将英语的被动语态译为汉语的无主句。在英语中，一些被动句是不能按照上述方法翻译的，如描述什么地方发生了什么样的事情，以及表示态度、观点、要求等的被动句，翻译时通常采用无主句，有时也要注意适当增译。

例如：Attention must be paid to safety in handling the flammable materials.

不当译文：注意安全应被放在处理易燃材料时。

正确译文：处理易燃材料时必须注意安全。

Exercise

Translate the following sentences into English.

1. Until recently, scientists were quite convinced that there wasn't a fifth force.
2. For this purpose, use is made of the difference between these two values.
3. Stress must be laid on the development of the electronics industry.

4. It is not known whether it is most often seen in inclusions of silicates such as garnet.
5. The process is similar to that performed at the transmit end, and the carriers are suppressed within the balanced modulation.
6. Theoretical analysis and design of thermal systems will not be directly dealt with.
7. Some C++ novices get irritated by the ambiguity reported by the compiler.
8. The air throughout a room becomes heated by convection currents.

Unit 6

Fluid Mechanics

Text A

Fluid Mechanics Fundamentals and Applications in Engineering

The Fundamentals of Fluid Mechanics

1. Fluid mechanics is a branch of physics concerned with the mechanics of fluids (liquids, gases, and plasmas) and the forces on them. Fluid mechanics has a wide range of applications, including mechanical engineering, civil engineering, chemical engineering, geophysics, astrophysics, and biology. Fluid mechanics can be divided into fluid statics, the study of fluids at rest; and fluid dynamics, the study of the effect of forces on fluid motion. It is a branch of continuum mechanics, a subject which models matter without using the information that it is made out of atoms; that is, it models matter from a macroscopic viewpoint rather than from microscopic. Fluid mechanics, especially fluid dynamics, is an active field of research with many problems that are partly or wholly unsolved. Fluid mechanics can be mathematically complex, and can best be solved by numerical methods, typically using computers. A modern **discipline**, called computational fluid dynamics (CFD)①, is devoted to this approach to solving fluid mechanics problems. Particle image velocimetry②, an experimental method for visualizing and analyzing fluid flow, also takes advantage of the highly visual nature of fluid flow.

---

① 计算流体动力学，是流体力学和计算机科学相互融合的一门新兴交叉学科。

② 粒子图像测速，是 20 世纪 70 年代末发展起来的一种瞬态、多点、无接触式的流体力学测速方法。

Fluid Statics and Fluid Dynamics

2. Fluid statics or **hydrostatics** is the branch of fluid mechanics that studies fluids at rest. It embraces the study of the conditions under which fluids are at rest in stable equilibrium, and is contrasted with fluid dynamics, the study of fluids in motion. Hydrostatics offers physical explanations for many phenomena of everyday life, such as why atmospheric pressure changes with altitude, why wood and oil float on water, and why the surface of water is always flat and horizontal whatever the shape of its container. Hydrostatics is fundamental to **hydraulics,** the engineering of equipment for storing, transporting and using fluids. It is also relevant to some aspect of geophysics and astrophysics (for example, in understanding plate **tectonics** and **anomalies** in the Earth's gravitational field), to **meteorology**, to medicine (in the context of blood pressure), and many other fields.

3. Fluid dynamics is a sub-discipline of fluid mechanics that deals with fluid flow—the science of liquids and gases in motion. Fluid dynamics offers a systematic structure—which underlies these practical disciplines—that embraces empirical and semi-empirical laws derived from flow measurement and used to solve practical problems. The solution to a fluid dynamics problem typically involves calculating various properties of the fluid, such as **velocity**, pressure, density, and temperature, as functions of space and time. It has several sub-disciplines itself, including **aerodynamics** (the study of air and other gases in motion) and **hydrodynamics** (the study of liquids in motion). Fluid dynamics has a wide range of applications, including calculating forces and moments on aircraft, determining the mass flow rate of petroleum through pipelines, predicting evolving weather patterns, understanding **nebulae** in **interstellar** space and modeling explosions. Some fluid-dynamical principles are used in traffic engineering and crowd dynamics.

Hydrodynamics and Aerodynamics

4. Hydrodynamics and aerodynamics are the specializations focusing on the motions of water and air, which are the most common fluids encountered in engineering. Those fields **encompass** not only the design of high-speed vehicles but also the motions of oceans and the atmosphere. Some engineers and scientists apply **sophisticated** computational models to **simulate** and understand interactions among the atmosphere, oceans, and global climates (Figure 6–1). The motion of fine pollutant particles in the air, improved weather forecasting, and the **precipitation** of raindrops and **hailstones** are some of the key issues that are addressed. The field of fluid mechanics is an exacting one, and many advances in it have occurred in conjunction with developments in applied mathematics and computer science. Fluids engineering fits within the broader context of the mechanical engineering topics shown in Figure 6–2.

Figure 6–1 The field of fluids engineering can involve the motion of fluids on very large—even **planetary**—scales.

Storms on the Earth, as well as the Great Red Spot[①] on Jupiter[②] shown here, form and move according to the principles of fluid mechanics.

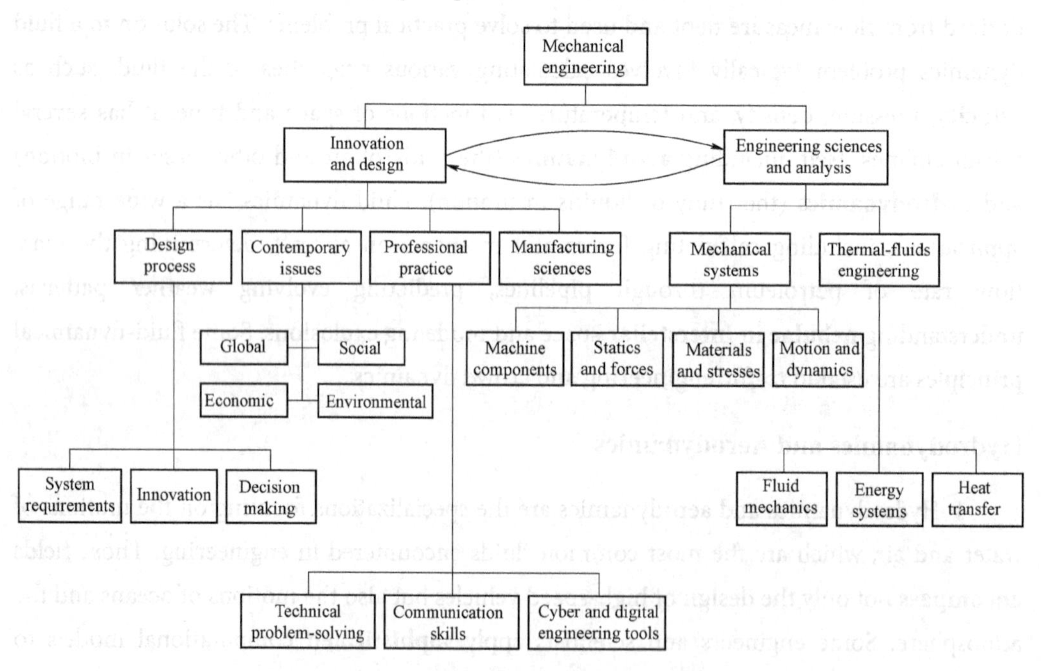

Figure 6–2 Relationship of the topics emphasized in this chapter relative to an overall program of study in mechanical engineering.

① "大红斑"，是一个反气旋旋涡，位于木星南赤道带边缘；"大红斑"面积广阔，宽度约等于地球圆周的3倍，其平均气温仅为–260℉。"大红斑"风暴，是整个太阳系最大的风暴，可能已存在了350年。

② 木星。

Fluid Mechanics Applications in Mechanical Engineering

5. Reflecting on the top-ten list of the mechanical engineering profession's achievements, 88% of the electricity in the America is produced by a process that involves continuously cycling water between liquid and steam, and back again. Coal, oil, natural gas, and nuclear fuels are used to heat water into steam, which in turn drives turbines and electrical generators. Another 7% of America's electricity is produced by hydroelectric power plants, and wind power provides a smaller fraction still. Collectively speaking, over 98% of the electricity in America is produced through processes that involve fluids engineering in one form or another.

6. The properties of fluids, the forces they generate, and the manner in which they flow from one location to another are key aspects of mechanical engineering. Fluid mechanics also plays a central role in biomedical engineering, a field that was ranked as one of the mechanical engineering profession's top ten achievements. Biomedical applications include the design of devices that deliver medicine by inhaling an aerosol spray and the flow of blood through **arteries** and **veins**. These devices are capable of performing chemical and medical **diagnostics** by exploiting the properties of fluids on microscopic scales. This emerging field, which is known as microfluidics, offers the potential for advances in **genomic** research and **pharmaceutical** discovery. Just as the field of electronics has undergone a revolution in **miniaturization**, chemical and medical laboratory equipment that presently fills an entire room is being miniaturized and made more economical.

7. The forces generated by **stationary** or moving fluids are important to the hardware designed by mechanical engineers. Forces are also generated by liquids and gases, and the fluid forces known as **buoyancy**, drag, and lift are researched by many scholars. Mechanical engineers apply sophisticated **computer-aided** engineering tools to understand complex air flows around aircraft and automobiles. In fact, those same methods have been applied to design golf balls capable of longer flight and to help ski jumpers, racing cyclists, marathon runners, and other athletes improve their performances.

(Source: https://basicmechanicalengineering.com/fluid-mechanics-fundamentals-and-applications-in-engineering/)

New Words and Expressions

artery	*n.*	动脉
aerosol	*adj.*	喷雾的
anomaly	*n.*	异常现象
aerodynamics	*n.*	气体力学

buoyancy	*n.*	浮力
computer-aided	*adj.*	计算机辅助的
diagnostics	*n.*	诊断学
discipline	*n.*	学科
encompass	*v.*	包含
hailstone	*n.*	冰雹
hydraulics	*n.*	水力学
hydrodynamics	*n.*	流体力学
hydrostatics	*n.*	流体静力学
interstellar	*adj.*	星际的
genomic	*adj.*	基因组的
tectonics	*n.*	构造地质学
meteorology	*n.*	气象学
miniaturization	*n.*	小型化
nebulae	*n.*	星云
pharmaceutical	*adj.*	制药的
planetary	*adj.*	行星的
precipitation	*n.*	降水量
stationary	*adj.*	静止的
sophisticated	*adj.*	复杂的
simulate	*v.*	模拟
velocity	*n.*	速度
vein	*n.*	静脉

Comprehension Checks

Work in pairs and discuss the following questions. Then share your ideas with the rest of the class.

1. What do you know about the major branches of fluid mechanics? (para.1)
2. In what way does hydrostatics help to explain some phenomena of daily life? State with an example. (para.2)
3. What fluid properties might be typically necessary in solving fluid dynamics problems? (para.3)
4. What are the key issues to be addressed when we model the interactions among atmosphere, ocean, and global climates? (para.4)
5. Could you briefly state some of the main duties of mechanical engineers? (para.6-7)

Vocabulary Tasks

I. Complete the sentences below. The first letter of the missing word has been given.

1. B________ is the ability that something has to float on a liquid or in the air.
2. The DNA sequencing and the data analysis will be compliant with regulations for lab d________.
3. If something e________ particular things, it includes them.
4. An i________ ship would be like an ark, carrying everything the colonists might need, including greenhouses for growing food and sophisticated manufacturing facilities.
5. A n________ is a cloud of dust and gas in space where new stars are produced.

II. Match each word with the correct definition.

1. artery	a. the study and use of systems that work using hydraulic pressure
2. hydraulics	b. rain, snow etc. that falls on the ground, or the amount of rain, snow etc. that falls
3. precipitation	c. the industrial production of medicines
4. sophisticated	d. small balls of ice that fall like rain from the sky
5. simulate	e. an important main route within a complex road, railway, or river system
6. pharmaceutical	f. the speed at which something moves in a particular direction
7. velocity	g. to make or produce something that is not real but has the appearance or feeling of being real
8. hailstone	h. more advanced or complex than others

III. Fill in the blanks with the words given below. Change the forms where necessary.

vein	miniaturization	genomic	aerosol	plasma
discipline	meteorology	planetary	stationary	anomaly

1. So far the search for longevity genes in humans has been extremely difficult, but prospects brighten as ________ technologies become faster and there are more centenarians to study.
2. The ________ noticed by Einstein kicked off one of the greatest revolutions in our understanding of the fabric of the universe.
3. Your ________ are the thin tubes in your body through which your blood flows toward your heart.

4. ________ is the study of the processes in the Earth's atmosphere that cause particular weather conditions, especially in order to predict the weather.
5. The improved structure is based on perpendicular magnetization and takes element ________ to below 30nm.
6. A(n) ________ can or spray is a small container in which a liquid such as paint or deodorant is kept under pressure.
7. Bolden said NASA's Mars research is the biggest part of the ________ science budget.
8. Blood cells like red blood cells float in the ________.
9. A ________ is a particular area of study, especially a subject of study in a college or university.
10. A burst of light is seen by two observers: one ________ on a platform, the other moving in a train.

Text B

1. Fluid dynamics is the study of the movement of fluids, including their interactions as two fluids come into contact with each other. In this context, the term "fluid" refers to either liquid or gases. It is a macroscopic, statistical approach to analyzing these interactions at a large scale, viewing the fluids as a **continuum** of matter and generally ignoring the fact that the liquid or gas is composed of individual atoms.

2. Fluid dynamics is one of the two main branches of fluid mechanics, with the other branch being fluid statics, the study of fluids at rest.

Basic Fluid Principles

3. The fluid concepts that apply in fluid statics also come into play when studying fluid that is in motion. Pretty much the earliest concept in fluid mechanics is that of buoyancy, discovered in ancient Greece by Archimedes①. As fluids flow, the density and pressure of the fluids are also crucial to understanding how they will interact. The **viscosity** determines how resistant the liquid is to change, so it is also essential in studying the movement of the liquid.

4. Here are some of the variables that come up in these analyses:

Bulk viscosity: μ

Density: ρ

Kinematic viscosity: $\nu = \mu / \rho$

① 阿基米德，古希腊哲学家、百科式科学家、数学家、物理学家、力学家，静态力学和流体静力学的奠基人。

Flow

5. Since fluid dynamics involves the study of the motion of fluid, one of the first concepts that must be understood is how physicists quantify that movement. The term that physicists use to describe the physical properties of the movement of liquid is flow.

6. Flow describes a wide range of fluid movement, such as blowing through the air, flowing through a pipe, or running along a surface. The flow of a fluid is classified in a variety of different ways, based upon the various properties of the flow.

Pipe Flow vs. Open-channel Flow

7. Pipe flow represents a flow that is in contact with rigid boundaries on all sides, such as water moving through a pipe or air moving through an air **duct**.

8. Open-channel flow describes flow in other situations where there is at least one free surface that is not in contact with a rigid boundary. In technical terms, the free surface has 0 parallel sheer stress. Cases of open-channel flow include water moving through a river, floods, water flowing during rain, tidal currents, and irrigation canals. In these cases, the surface of the flowing water, where the water is in contact with the air, represents the "free surface" of the flow.

9. Flows in a pipe are driven by either pressure or gravity, but flows in open-channel situations are driven solely by gravity.

10. City water systems often use water towers to take advantage of this, so that the *elevation* difference of the water in the tower (the hydrodynamic head) creates a differential pressure, which are then adjusted with mechanical pumps to get water to the locations in the system where they are needed.

Application of Fluid Dynamics

11. Two-thirds of the Earth's surface is water and the planet is surrounded by layers of atmosphere, so we are literally surrounded at all times by fluids. Thinking about it for a bit, this makes it pretty obvious that there would be a lot of interactions of moving fluids for us to study and understand scientifically. That's where fluid dynamics comes in, so there is no shortage of fields that apply concepts from fluid dynamics.

12. This list is not at all exhaustive, but provides a good overview of ways in which fluid dynamics shows up in the study of physics across a range of specializations.

13. *Oceanography, Meteorology & Climate Science* — Since the atmosphere is modeled as fluids, the study of atmospheric science and ocean currents, crucial for understanding and predicting weather patterns and climate trends, relies heavily on fluid dynamics.

14. ***Aeronautics*** — The physics of fluid dynamics involves studying the flow of air to

create drag and lift, which in turn generate the forces that allow heavier-than-air flight.

15. *Geology & Geophysics* — Plate tectonics involves studying the motion of the heated matter within the liquid core of the Earth.

16. ***Hematology*** & ***Hemodynamics*** — The biological study of blood includes the study of its circulation through blood **vessels**, and the blood circulation can be modeled using the methods of fluid dynamics.

Alternative Names of Fluid Dynamics

17. Fluid dynamics is also sometimes referred to as hydrodynamics, although this is more of a historical term. Throughout the twentieth century, the phrase "fluid dynamics" became much more commonly used. Technically, it would be more appropriate to say that hydrodynamics is when fluid dynamics is applied to liquids in motion and aerodynamics is when fluid dynamics is applied to gases in motion. However, in practice, specialized topics such as hydrodynamic stability and **magnetohydrodynamics** use the "hydro-" prefix even when they are applying those concepts to the motion of gases.

(Source: https://www.thoughtco.com/what-is-fluid-dynamics-4019111)

New Words and Expressions

aeronautics	*n.*	航空学
bulk	*n.*	体积，容量
continuum	*n.*	连续统一体
duct	*n.*	输送管
hematology	*n.*	血液学
hemodynamics	*n.*	血液动力学
kinematic	*adj.*	运动学上的
magnetohydrodynamics	*n.*	磁动流体力学
viscosity	*n.*	黏度

Comprehension Questions

1. According to the passage, which of the following statements is TRUE about fluid dynamics?

A. It is a microscopic statics analyzing the interactions between two fluids.

B. In the study of fluid dynamics, it should be taken into consideration that the liquid or gas is composed of individual atoms.

C. Fluid dynamics is the subject studying the movement of fluid which typically refers to liquid or gases.

2. Which of the following fluid movement should NOT be considered as “flow”?
   A. Blowing through the air.
   B. Boiling in a kettle.
   C. Running along a surface.
3. What does the underlined word “elevation” (Line 1, para. 10) probably mean?
   A. 高度　　B. 压力　　C. 重量
4. What can we learn from the application of fluid dynamics in hematology and hemodynamics?
   A. The physics of fluid dynamics involves studying the flow of air to create drag and lift, which in turn generates the forces that allow heavier-than-air flight.
   B. The blood circulation can be modeled using the methods of fluid dynamics.
   C. The whole body of traffic can be treated as a single entity that behaves in ways roughly similar to the flow of a fluid.
5. Fluid dynamics can also be referred to as ________.
   A. hydrodynamics
   B. hemodynamics
   C. magnetohydrodynamics

rammar Coach

Non-finite Verb

非谓语动词（non-finite verb），又叫非限定动词，是指在句子中不作谓语的动词，不受主语的人称和数的限制。非谓语动词有四种常见形式：动名词、动词不定式、现在分词和过去分词。动词的非谓语形式可以替代名词性从句，定语从句或状语从句，并可充当主语、宾语、主语补足语、定语及其他成分。

1. 科技英语中非谓语动词的特点

（1）动名词的特点。动名词是由动词后加上-ing 的词尾构成的。由于动名词是由动词变化而来的，具有动词的一些特征和动词的变化形式，因此可以用来表达名词所不能够完全表达的意思；同时，动名词也具有名词的特征，动名词及其短语可充当句子的主语、表语、宾语、主语补足语与宾语补足语以及定语等。

例如： With a magnitude of 8.1, the tremor was capable of destroying entire towns.

本句中的非谓语动词 destroying 具有名词的特征，同时也具有动词的特征，充当介词 of 的宾语。

（2）动词不定式的特点。动词不定式是由“to+动词原形”构成的一种非谓语动词结构，其否定形式是“no/never to +动词原形”的形式。动词不定式的一般形式 to do 表示将要发生的具体动作；to be done为不定式被动式，体现其逻辑主语是动作的承受者，而非动作的执行者。不定式可代替名词和形容词，在句子中可作主语、宾语、表语和定语。

例如：Several neutrons are also released which can go on to split other nearby atoms, producing a nuclear chain reaction of sustained energy release.

本句中动词不定式to spilt作状语，表示目的。

（3）分词的特点。分词是动词的非限定形式，其中现在分词由“动词原形+ing”构成。

例如：A direct current is a current flowing always in the same direction.

过去分词有规则动词和不规则动词的过去分词之分，规则动词的过去分词是由“动词+-ed”构成，不规则动词则有各自不同的过去分词形式，如kept、broken、shut、written等。

例如：The nuclear waste can be handled properly and disposed of geologically without affecting the environment in any way.

本句中，handled、disposed就分别是handle、dispose的过去分词形式。

作为非谓语动词形式，分词的作用相当于形容词和副词，但它们仍然保留动词的基本特征，即可以有自己的宾语和状语。分词与其宾语、状语一起构成分词短语。分词和分词短语都可以在句子中充当定语、状语、表语、宾语补足语。例如：

① The generative CAPP system synthesizes optimum process sequence, based on an analysis of part geometry, material, and other factors which would influence manufacturing decisions.（分词作定语）

② Before starting any check of operation, confirm that the cover is mounted on the high-voltage section.（分词作状语）

③ We put a hand above an electric fire and feel the hot air rising.（分词作宾语补足语）

④ Standard NC machines greatly reduced the machining time required to produce a part or complete a production run of parts, but the overall operation was still time-consuming.（分词作表语）

⑤ Other things being equal, a diamond with a higher clarity-grade has a higher value.（分词的独立主格结构）

2. 科技英语中非谓语动词的翻译策略

（1）非谓语动词译为名词。从词汇意义方面来看，非谓语动词可分为状态动词和动作动词，状态动词并非表示动作，而是表示一种较为静止的状态。对于这类表示主语特征或状态的非谓语动词，为了适应汉语的表达习惯，可转译为名词。

例如：Optimizing heating and cooling appliances is another major way to improve the energy performance of most homes.

不当译文：优化供暖的电器和制冷的电器是改善多数家庭能源性能的重要方式。

正确译文：优化供暖和制冷电器是改善多数家庭能源性能的重要方式。

解析：原句中的非谓动词转译为汉语中的名词，在不改变原词词义的前提下，使译文更加简洁，符合汉语表达方式。

（2）非谓语动词译为动词。非谓语动词是由动词转变而来，因此它在词义和用法方面都有动词的特点。汉语中并不存在非谓语动词，因此在翻译的过程中，大多数的非谓语动词可直接译为汉语中的动词。

例如：The best way of lowering the cost of solar energy is to improve the cell's efficiency.

不当译文：太阳能成本降低的最佳方式是来提升电池的效率。

正确译文：降低太阳能成本的最佳方式是提升电池效率。

解析：原句中存在两个非谓语动词 lowering 和 to improve，lowering 译为汉语动词"降低"，to improve 译为汉语动词"提升"，既体现了译文与原文在词汇方面的动态对等，也符合汉语多用动词的习惯。

（3）非谓语动词译为形容词。非谓语动词在句子中起修饰名词的作用时，多译为形容词放在名词前。非谓语动词译为形容词的情况主要有三种：① 单个现在分词或过去分词作定语；② 单个现在分词或过去分词作表语；③ 不定式置于被修饰的名词之后。

例如：Only in the last few decades when growing energy demands, increasing environmental problems and declining fossil fuel resources made us look to alternative energy options have we focused our attention on truly exploiting this tremendous resources.

不当译文：仅在过去几十年里，能源需求增长、环境问题增加，以及化石燃料减少，使得人们开始寻找替代能源，并且致力于开发这种丰富的资源。

正确译文：仅在过去几十年里，随着不断增长的能源需求、日益严重的环境问题，以及日渐枯竭的化石燃料问题的出现，人们开始寻找可替代能源，并且致力于开发这种丰富的资源。

解析：原句中画线的三个动名词，即 growing、increasing、declining，分别修饰 energy demands、environmental problems、fossil fuel resources，以凸显问题的严重性，都可以译为形容词放在名词前面作定语，使得语言搭配更为地道。

Exercise

Translate the following sentences into Chinese and pay attention to the non-finite verbs.

1. There are different ways of changing energy from one form into another.
2. Radiating from the Earth, heat causes air currents to rise.
3. The existence of the giant clouds was virtually required for the Big Bang, first put forward in the 1920s, to maintain its reign as the dominant explanation of the cosmos.
4. Materials to be used for structural purposes are chosen so as to behave elastically in the environmental conditions.
5. Some of these causes are completely reasonable results of social needs. Others are

reasonable consequences of particular advances in science <u>being</u> to some extent self-accelerating.

6. The web interface also provides some of the logistical data <u>mentioned</u> in the previous section, and can alert the user via e-mail in the event of the failure.

Unit 7

Heat Transfer

Text A

Heat and Temperature

What is Heat?

1. Consider a very hot mug of coffee on the **countertop** of your kitchen. For discussion purposes, we will say that the cup of coffee has a temperature of 80°C and that the surroundings (countertop, air in the kitchen, etc.) has a temperature of 26°C. What do you suppose will happen in this situation? I suspect that you know that the cup of coffee will gradually cool down over time. At 80°C, you wouldn't dare drink the coffee. Even the mug will likely be too hot to touch. But over time, both the mug and the coffee will cool down. Soon it will be at a drinkable temperature. And if you resist the **temptation** to drink the coffee, it will eventually reach room temperature. The coffee cools from 80°C to about 26°C. So what is happening over the course of time to cause the coffee to cool down? The answer to this question can be both **macroscopic** and **particulate** in nature.

2. On the macroscopic level, we would say that the coffee and the mug are transferring heat to the surroundings. This transfer of heat occurs from the hot coffee and hot mug to the surrounding air. The fact that the coffee lowers its temperature is a sign that the average **kinetic** energy[①] of its **particles** is decreasing. The coffee is losing energy. The mug is also lowering its temperature; the average kinetic energy of its particles is also decreasing. The mug is also losing energy. The energy that is lost by the coffee and the mug is being transferred to the colder surroundings. We refer to this transfer of energy from the coffee and

① 动能，即物体由于运动而具有的能量，为物体质量与速度平方乘积的一半。

the mug to the surrounding air and countertop as heat. In this sense, heat is simply the transfer of energy from a hot object to a colder object.

3. Now let's consider a different scenario — that of a cold can of pop placed on the same kitchen counter top. For discussion purposes, we will say that the pop and the can which contains it has a temperature of 5°C and that the surroundings (countertop, air in the kitchen, etc.) has a temperature of 26°C. What will happen to the cold can of pop over the course of time? Once more, I suspect that you know the answer. The cold pop and the container will both warm up to room temperature. But what is happening to cause these colder-than-room-temperature objects to increase their temperature? Is the cold escaping from the pop and its container? No! There is no such thing as the cold escaping or leaking. Rather, our explanation is very similar to the explanation used to explain why the coffee cools down. There is a heat transfer①.

4. Over time, the pop and the container increase their temperature. The temperature rises from 5°C to nearly 26°C. This increase in temperature is a sign that the average kinetic energy of the particles within the pop and the container is increasing. In order for the particles within the pop and the container to increase their kinetic energy, they must be gaining energy from somewhere. But from where? Energy is being transferred from the surroundings (countertop, air in the kitchen, etc.) in the form of heat. Just as in the case of the cooling coffee mug, energy is being transferred from the higher temperature objects to the lower temperature object. Once more, this is known as heat — the transfer of energy from the higher temperature object to a lower temperature object, as shown in Figure 7–1.

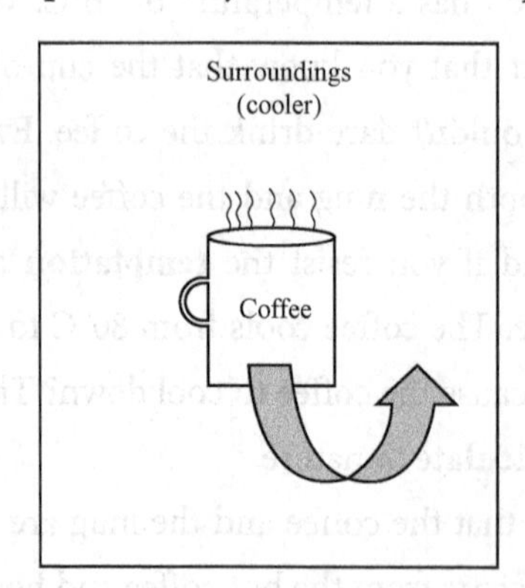

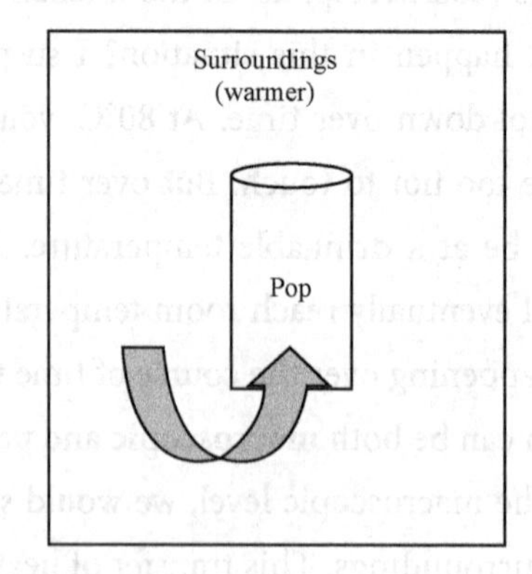

Figure 7–1 Heat

Another Definition of Temperature

5. Both of these scenarios could be summarized by two simple statements. An object decreases its temperature by releasing energy in the form of heat to its surroundings. And an

① 热传递，是指存在温度差的条件下，热量从一个物体转移到另一个物体或从物体的一部分转移到另一部分的过程。

object increases its temperature by gaining energy in the form of heat from its surroundings. Both the warming up and the cooling down of objects works in the same way — by heat transfer from the higher temperature object to the lower temperature object. So now we can meaningfully re-state the definition of temperature. Temperature is a measure of the ability of a substance, or more generally of any physical system, to transfer heat to another physical system. The higher the temperature of an object is, the greater the tendency of that object to transfer heat. The lower the temperature of an object is, the greater the tendency of that object to be on the receiving end of the heat transfer.

6. But perhaps you have been asking: what happens to the temperature of surroundings? Do the countertop and the air in the kitchen increase their temperature when the mug and the coffee cool down? And do the countertop and the air in the kitchen decrease its temperature when the can and its pop warm up? The answer is a resounding "Yes"! The proof? Just touch the countertop — it should feel cooler or warmer than before the coffee mug or pop can were placed on the countertop. But what about the air in the kitchen? Now that's a little more difficult to present a convincing proof of the decrease of the kitchen air. The fact that the **volume** of air in the room is so large and that the energy quickly **diffuses** away from the surface of the mug means that the temperature change of the air in the kitchen will be abnormally small. In fact, it will be **negligibly** small. There would have to be a lot more heat transfer before there is a noticeable temperature change.

Thermal Equilibrium[①]

7. In the discussion of the cooling of the coffee mug, the countertop and the air in the kitchen were referred to as the surroundings. It is common in physics discussions of this type to use a mental framework of a system and the surroundings. The coffee mug and the coffee would be regarded as the system and everything else in the universe would be regarded as the surroundings. To keep it simple, we often narrow the scope of the surroundings from the rest of the universe to simplify those objects that are immediately surrounding the system.

8. Now let's imagine a third situation. Suppose that a small metal cup of hot water is placed inside of a larger **Styrofoam** cup of cold water. Let's suppose that the temperature of the hot water is **initially** 70°C and that the temperature of the cold water in the outer cup is initially 5°C. And let's suppose that both cups are equipped with thermometers that measure the temperature of the water in each cup over the course of time. What do you suppose will happen? Before you read on, think about the question and commit to some form of answer. When the cold water is done warming and the hot water is done cooling, will their temperatures be the same or different? Will the cold water warm up to a lower temperature

① 热平衡，是指同外界接触的物体其内部温度各处均匀且等于外界温度的状况。

than the temperature that the hot water cools down to? Or as the warming and cooling occurs, will their temperatures cross each other?

9. Fortunately, this is an experiment that can be done and in fact has been done on many occasions. Figure 7–2 is a typical representation of the results.

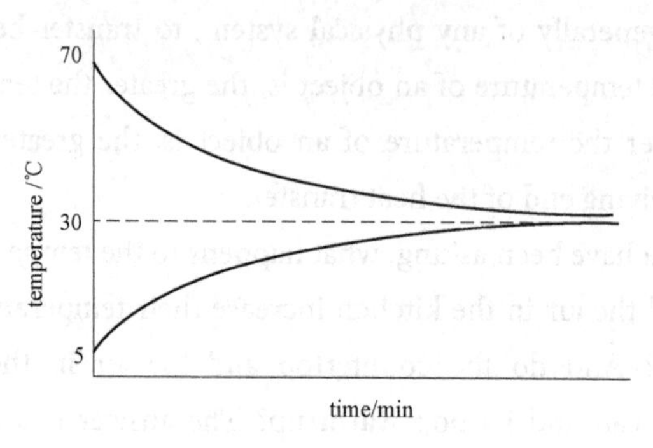

Figure 7-2 temperature varition over time

10. As you can see from the graph, the hot water cooled down to **approximately** 30°C and the cold water warmed up to approximately the same temperature. Heat is transferred from the high temperature object (inner can of hot water) to the low temperature object (outer can of cold water). If we **designate** the inner cup of hot water as the system, then we can say that there is a flow of heat from the system to the surroundings. As long as there is a temperature difference between the system and the surroundings, there is a heat flow between them. The heat flow is more rapid at first as depicted by the steeper slopes of the lines. Over time, the temperature difference between system and surroundings decreases and the rate of heat transfer decreases. This is denoted by the gentler slope of the two lines. Eventually, the system and the surroundings reach the same temperature and the heat transfer ceases. It is at this point, that the two objects are said to have reached thermal equilibrium.

(Source: http://www.physicsclassroom.com/class/thermalP/Lesson-1/What-is-Heat)

New Words and Expressions

approximately	*adv.*	近似地
countertop	*n.*	（厨房的）工作台面
designate	*v.*	指定
diffuse	*v.*	传播，四散
initially	*adv.*	最初，开始
kinetic	*adj.*	运动的，能动的
macroscopic	*adj.*	宏观的

negligibly *adv.* 微不足道地
particle *n.* 微粒，颗粒
particulate *adj.* 微粒的，粒子的
styrofoam *n.* 泡沫塑料，发泡胶
temptation *n.* 诱惑，引诱
volume *n.* 体积，大量

Comprehension Checks

Work in pairs and discuss the following questions. Then share your ideas with the rest of the class.

1. What is happening when the colder-than-room-temperature objects increase their temperature? (para.3)
2. What are the necessary conditions for heat transfer?
3. Would you sense the decrease of temperature of kitchen air when the can and its pop warm up? (para. 6)
4. At what point can two objects reach thermal equilibrium? (para.10)

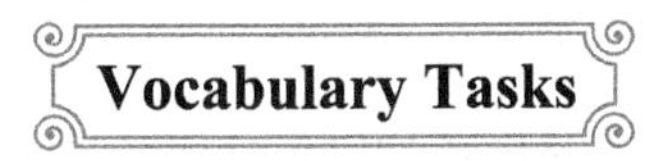

Vocabulary Tasks

I. Complete the sentences below. The first letter of the missing word has been given.

1. T________ is a measure of the ability of a substance, or more generally of any physical system, to transfer heat energy to another physical system.
2. The v_______ of air in the room is so large and that the energy quickly d_______ away from the surface of the mug.
3. The hot water cooled down to a________ 30°C and the cold water warmed up to about the same temperature.
4. Let's suppose that both cups are equipped with t________ that measure the temperature of the water in each cup over the course of time.
5. Eventually, the system and the surroundings reach the same temperature and the heat t________ ceases.

II. Match each word with the correct definition.

1. heat — a. relating to or caused by heat or by changes in temperature
2. thermal — b. visible to the naked eye
3. diffuse — c. the act of moving something from one location to another
4. equilibrium — d. the mechanical energy that a body has by virtue of its motion

5. volume	e. the quality of being hot
6. macroscopic	f. to spread something in all directions
7. kinetic energy	g. balance between several different influences or aspects of a situation
8. transfer	h. the amount of something or the amount of space that an object or a substance fills

III. Fill in the blanks with the words given below. Change the forms where necessary.

heat	contain	temperature	substance	approximately
transfer	cease	temptation	volume	surrounding

1. The fact shows that the average unit of energy consumption is ________ 300 k•Wh/t by using blowing quick freezing to process food.
2. Having spent so long at a great club like Rangers, therefore, no other Scottish team could ________ him away.
3. The seas store ________ and release it gradually during cold periods.
4. The characteristics of novel ________ port are energy saving, smart and efficient.
5. Some nodes ________ working before receiving all new programs due to the shortage of energy.
6. Normally, helium is present in less than 1% by ________ in natural gas.
7. The lead poisoning rate of children living in ________ mine areas is significantly higher than that of the urban children and national average.
8. The presence or absence of clouds can have an important impact on heat ________.
9. It is found that the main chemical constituents of the scale are organic ________.
10. It is proved that during the process of outburst, the rise of coal ________ is due to the release of elastic energy while the coal is cracked by crustal stress.

Text B

Methods of Heat Transfer

Conduction — A Particle View

1. At the macroscopic level, heat is the transfer of energy from the high temperature object to the low temperature object. At the particle level, heat flow can be explained in terms of the net effect of the **collisions** of a whole bunch of little bangers. Warming and cooling is the macroscopic result of this particle-level phenomenon. Now let's apply this particle view to

the scenario of the metal can with the hot water positioned inside of a Styrofoam cup containing cold water. On average, the particles with the greatest kinetic energy are the particles of the hot water. Being a fluid, those particles move about with **translational** kinetic energy (abbr. KE) and bang upon the particles of the metal can. As the hot water particles bang upon the particles of the metal can, they transfer energy to the metal can. This warms the metal can up. Most metals are good thermal **conductors** so they warm up quite quickly throughout the bulk of the can. The can assumes nearly the same temperature as the hot water. Being a solid, the metal can consists of little wigglers. The wigglers at the outer **perimeter** of the metal can bang upon particles in the cold water. The collisions between the particles of the metal can and the particles of the cold water result in the transfer of energy to the cold water. This slowly warms the cold water up. The interaction between the particles of the hot water, the metal can and the cold water results in a transfer of energy outward from the hot water to the cold water. The average kinetic energy of the hot water particles gradually decreases; the average kinetic energy of the cold-water particles gradually increases; and eventually, thermal equilibrium would be reached at the point that the particles of the hot water and the cold water have the same average kinetic energy, as shown in Figure 7–3. At the macroscopic level, one would observe a decrease in temperature of the hot water and an increase in temperature of the cold water.

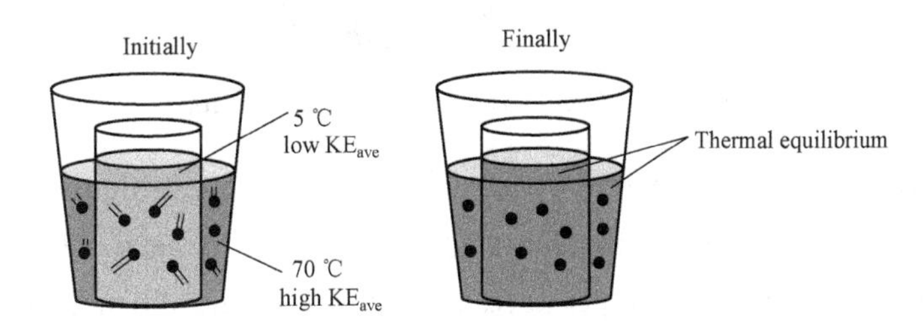

Figure 7–3 Thermal equilibrium [The average KE (KE_{ave}) and the temperature of each sample of water are equal.]

2. The **mechanism** in which heat is transferred from one object to another object through particle collisions is known as **conduction.** In conduction, there is no net transfer of physical stuff between the objects. Nothing material moves across the boundary. The changes in temperature are wholly explained as the result of the gains and losses of kinetic energy during collisions.

Conduction through the Bulk of an Object

3. The mechanism of heat transfer through the bulk of the **ceramic** mug is described in a similar manner as before. The ceramic mug consists of a collection of orderly arranged wigglers. These are particles that **wiggle** about a fixed position. As the ceramic particles at the

boundary between the hot coffee and the mug warm up, they attain a kinetic energy that is much higher than their neighbors. As they wiggle more **vigorously**, they bang into their neighbors and increase their vibrational kinetic energy as shown in Figure 7–4. These particles in turn begin to wiggle more vigorously and their collisions with their neighbors increase their vibrational kinetic energy. The process of energy transfer by means of the little bangers continues from the particles at the inside of the mug (in contact with the coffee particles) to the outside of the mug (in contact with the surrounding air). Soon the entire coffee mug is warm and your hand feels it.

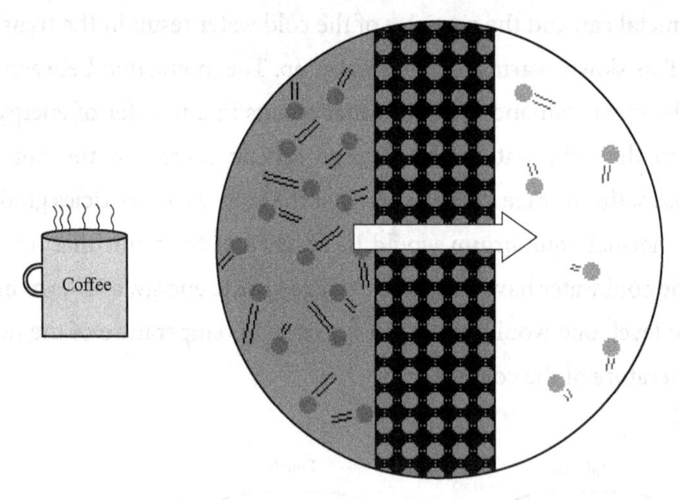

Figure 7–4 Conduction through a ceramic mug

4. This mechanism of conduction by particle-to-particle interaction is very common in ceramic materials such as a coffee mug. Does it work the same in metal objects? For instance, you likely have noticed the high temperatures attained by the metal handle of a **skillet** when placed upon a stovetop. The burners on the stove transfer heat to the metal skillet. If the handle of the skillet is metallic, it too attains a high temperature, certainly high enough to cause a bad burn. The transfer of heat from the skillet to the skillet handle occurs by conduction. But in metals, the conduction mechanism is slightly more complicated. In a manner similar to electrical conductivity, thermal conductivity in metals occurs by the movement of free **electrons**. Outer shell electrons of metal atoms are shared among atoms and are free to move throughout the bulk of the metal. These electrons carry the energy from the skillet to the skillet handle. The details of this mechanism of thermal conduction in metals are considerably more complex than the discussion given here. The main point to grasp is that heat transfer through metals occurs without any movement of atoms from the skillet to the skillet handle. This qualifies the heat transfer as being categorized as thermal conduction.

Heat Transfer by Convection

5. **Convection**① is the process of heat transfer from one location to the next by the movement of fluids. The moving fluid carries energy with it. The fluid flows from a high temperature location to a low temperature location.

6. To understand convection in fluids, let's consider the heat transfer through the water that is being heated in a pot on a stove. Of course the source of the heat is the stove burner. The metal pot that holds the water is heated by the stove burner. As the metal becomes hot, it begins to conduct heat to the water. The water at the boundary with the metal pan becomes hot. Fluids expand when heated and become less dense. So as the water at the bottom of the pot becomes hot, its density decreases. Differences in water density between the bottom of the pot and the top of the pot results in the gradual formation of circulation currents. Hot water begins to rise to the top of the pot displacing the colder water that was originally there. And the colder water that was present at the top of the pot moves towards the bottom of the pot where it is heated and begins to rise. These circulation currents slowly develop over time, providing the pathway for heated water to transfer energy from the bottom of the pot to the surface as shown in Figure 7–5.

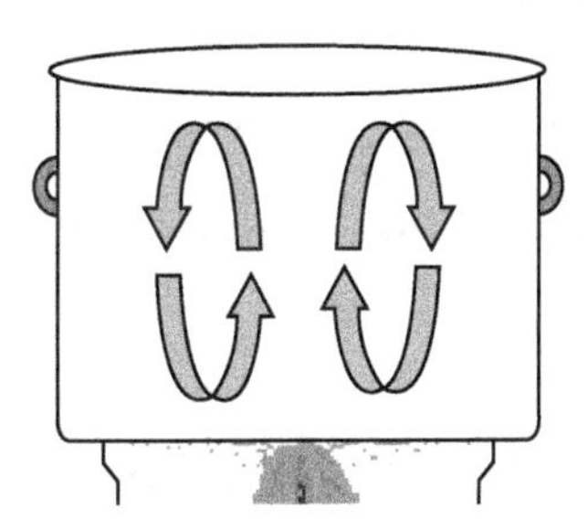

Figure 7–5 Convection in fluids

7. Convection also explains how an electric heater placed on the floor of a cold room warms up the air in the room. Air present near the coils of the heater warm up. As the air warms up, it expands, becomes less dense and begins to rise. As the hot air rises, it pushes some of the cold air near the top of the room out of the way. The cold air moves towards the bottom of the room to replace the hot air that has risen. As the colder air approaches the heater at the bottom of the room, it becomes warmed by the heater and begins to rise. Once more, convection currents are slowly formed. Air travels along these pathways, carrying energy with it from the heater throughout the room.

8. Convection is the main method of heat transfer in fluids such as water and air. It is often said that heat rises in these situations. The more appropriate explanation is to say that

① 对流，流体各部分之间发生相对位移、依靠冷热流体互相掺混和移动所引起的热量传递方式。

heated fluid rises. For instance, as the heated air rises from the heater on a floor, it carries more energetic particles with it. As the more energetic particles of the heated air mix with the cooler air near the ceiling, the average kinetic energy of the air near the top of the room increases. This increase in the average kinetic energy corresponds to an increase in temperature. The net result of the rising hot fluid is the transfer of heat from one location to another location. The convection method of heat transfer always involves the transfer of heat by the movement of matter. Our model of convection considers heat to be energy transfer that is simply the result of the movement of more energetic particles.

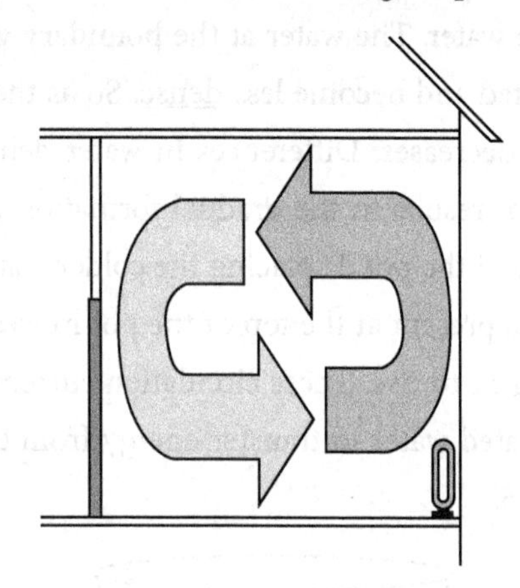

Figure 7–6 Convection in air

(Source: http://www.physicsclassroom.com/class/thermalP/Lesson-1/Methods-of-Heat-Transfer)

New Words and Expressions

ceramic	*adj.*	陶瓷的
collision	*n.*	碰撞，冲突
conduction	*n.*	（热、电等的）传导
conductors	*n.*	（电）导体
convection	*n.*	对流，传送
electron	*n.*	电子
mechanism	*n.*	机制，机理
perimeter	*n.*	周围，边界
skillet	*n.*	平底煎锅
translational	*adj.*	平动的，平移的
vigorously	*adv.*	剧烈地
wiggle	*v.*	摆动，扭动

Comprehension Questions

1. Which of the following statements is TRUE about the scenario mentioned in para.1?
   A. Particles of the cold water have greater kinetic energy than particles of the hot water.
   B. At the particle-level, when the particles of the hot water and the cold water reach the same temperature, thermal equilibrium will be achieved.
   C. When particles of hot water and the metal can collide with each other, the energy is transferred to the metal can.
2. Which of the following statements is WRONG about the conduction mechanism in metals? (para.4)
   A. Atoms share the outer electrons of metals which move freely throughout the metal bulk.
   B. The motion of free electrons causes the thermal conductivity in metals.
   C. In heat transfer through metals, atoms move from the skillet to the skillet handle.
3. What does the underlined word "dense" (line 5, para. 6) probably mean?
   A. 稀疏的　　　B. 稠密的　　　C. 稳定的
4. The major difference between conduction and convection lies in ________.
   A. whether the high temperature location and the low temperature location will change in temperature
   B. whether there are gains and losses of energy between the high temperature and the low temperature location
   C. whether it is the main method of heat transfer in fluids or in solids
5. Which of the following correctly describes the heat transfer by convection?
   A. It always involves the transfer of heat by the movement of matter.
   B. It happens when heat rises in solids.
   C. It happens when fluids flow from a low temperature location to a high temperature location.

Post-positive Attributive

后置定语（post-positive attributive）就是置于名词之后，用来修饰名词的词。形容词（短语）、介词短语、分词、定语从句等都可以充当后置定语。科技英文文体严谨周密，行文简练，重点突出。因为其具有抽象性、专业性和理论性等特点，所以会大量使用后置定语表达一些复杂的概念和严密的逻辑关系。以下将通过举例介绍形容词（短语）、介词短语、分词及定语从句作后置定语的语法现象。

1. 形容词及形容词短语作后置定语

例如：

（1）We need a timing method **accurate to millionths of a second.**

accurate to millionths of a second（精确到百万分之几秒的）为形容词短语作后置定语，由于定语过长，所以选择将其后置，既强调了 timing method 的特性，也避免了句子“头重脚轻”。

（2）If adopting some connection methods, **neither series nor parallel**, the above phenomenon will occur.

neither series nor parallel（既非串联也非并联的）为形容词短语作后置定语，修饰 connection methods（连接方式），强调了连接方式的特性。

2. 介词短语作后置定语

例如：

（1）The zone theory has held up well as we continue to learn more **about the coal seam.**

about the coal seam（关于煤层的）为介词 about 引导的短语作后置定语，修饰 more。具体说明了学习的主要内容是和煤层相关的。

（2）The forces **due to friction** are called friction forces.

due to friction（由于摩擦的原因）为介词短语作 forces（摩擦力）的后置定语，说明了摩擦力产生的原因。

3. 分词作后置定语

例如：

（1）All cars **passing by** are driven by electricity.

passing by（刚刚经过的车辆）为现在分词作 cars 的后置定语，既说明了所指车的特点，即刚经过的，又强调了进行的动作。

（2）The building **built** last month needs repairing.

built last month（上个月修建的）为过去分词作 building 的后置定语，既表示被动（即这座大楼是被修建的），又表示完成（即这座大楼上个月就已经修建完成）。

4. 定语从句作后置定语

例如：

（1）The sun **which heats the earth** makes plants grow.

which heats the earth（照耀大地的）为关系代词 which 引导的定语从句作后置定语，修饰 sun，说明了正是因为太阳的特性，即能够温暖大地，才使得植物生长。

（2）Blackbody radiators are light sources **whose photons are purely the result of thermal energy given off by atoms.**

whose photons are purely the result of thermal energy given off by atoms 是由关系副词 whose 引导的定语从句，修饰 light resources（光源体），用定语从句具体地解释了光源

体的特性和产生原理，既全面，又避免了整个句子“头重脚轻”。

Exercise

Translate the following sentences into English.

1. 工厂唯一可用的燃料是煤。（形容词作后置定语）
2. 在网上购买的日常用品价格低于在商店的价格。（分词作后置定语）
3. 这是一种由 6 个部分构成的设备。（分词作后置定语）
4. 新能源产业的发展有助于能源的稳定发展。（介词短语作后置定语）
5. 该工厂所采用的经济发展方式既环保又利于可持续发展。（定语从句作后置定语）

Unit 8

Mechanical Engineering

Text A

What is Mechanical Engineering?

1. Mechanical engineering is the discipline that applies engineering, physics, and materials science principles to design, analyze, manufacture, and maintain mechanical systems. It is the branch of engineering that involves the design, production, and operation of machinery. It is one of the oldest and broadest of the engineering disciplines.

2. The mechanical engineering field requires an understanding of core areas including mechanics, **kinematics, thermodynamics**, materials science, structural analysis, and electricity. In addition to these core principles, mechanical engineers use tools such as computer-aided design (CAD), and product life cycle management to design and analyze manufacturing plants, industrial equipment and machinery, heating and cooling systems, transport systems, aircraft, watercraft, robotics, medical devices, weapons, and others.

3. Mechanical engineering emerged as a field during the Industrial Revolution in Europe in the 18th century; however, its development can be traced back several thousand years around the world. In the 19th century, developments in physics led to the development of mechanical engineering science. The field has continually evolved to incorporate advancements; today mechanical engineers are pursuing developments in such areas as **composites, mechatronics**, and **nanotechnology**. It also overlaps with aerospace engineering, **metallurgical** engineering, civil engineering, electrical engineering, manufacturing engineering, chemical engineering, industrial engineering, and other engineering disciplines to varying amounts. Mechanical engineers may also work in the field of biomedical engineering, specifically with **biomechanics**, transport phenomena, **biomechatronics**,

bionanotechnology, and modelling of biological systems.

Modern Tools

4. Many mechanical engineering companies, especially those in industrialized nations, have begun to incorporate computer-aided engineering (CAE) ① programs into their existing design and analysis processes, including 2D and 3D solid modeling computer-aided design (CAD). This method has many benefits, including easier and more exhaustive visualization of products, the ability to create virtual assemblies of parts, and the ease of use in designing mating interfaces and tolerances.

5. Other CAE programs commonly used by mechanical engineers include product lifecycle management (PLM) tools and analysis tools used to perform complex **simulations**. Analysis tools may be used to predict product response to expected loads, including fatigue life and manufacturability. These tools include finite element analysis (FEA) ②, computational fluid mechanics (CFM) ③, and computer-aided manufacturing (CAM) ④.

6. Using CAE programs, a mechanical design team can quickly and cheaply **iterate** the design process to develop a product that better meets cost, performance, and other constraints. No physical **prototype** needs be created until the design nears completion, allowing hundreds or thousands of designs to be evaluated, instead of a relative few. In addition, CAE analysis programs can model complicated physical phenomena which cannot be solved by hand, such as **viscoelasticity**, complex contact between mating parts, or non-Newtonian flows.

7. As mechanical engineering begins to merge with other disciplines, as seen in mechatronics, multidisciplinary design optimization (MDO) is being used with other CAE programs to automate and improve the iterative design process. MDO tools wrap around existing CAE processes, allowing product evaluation to continue even after the analyst goes home for the day. They also utilize sophisticated optimization **algorithms** to more intelligently explore possible designs, often finding better, innovative solutions to difficult multidisciplinary design problems.

Subdisciplines

8. The field of mechanical engineering can be thought of as a collection of many mechanical engineering science disciplines. Several of these subdisciplines which are typically taught at the undergraduate level are listed below, with a brief explanation and

---

① 计算机辅助工程，该工程利用计算机辅助求解分析复杂工程和产品的结构力学性能，以及优化结构性能等，把工程（生产）的各个环节有机地组织起来。

② 有限元分析，其基本思想是用较简单的问题代替复杂问题后再求解。

③ 计算流体力学，指通过计算机和数值方法来求解流体力学的控制方程，并对流体力学问题进行模拟和分析。

④ 计算机辅助制造，指利用计算机辅助完成从生产准备到产品制造整个过程的活动。

the most common application of each. Some of these subdisciplines are unique to mechanical engineering, while others are a combination of mechanical engineering and one or more other disciplines. Most work that a mechanical engineer does uses skills and techniques from several of these subdisciplines, as well as specialized subdisciplines.

Mechanics

9. Mechanics is, in the most general sense, the study of forces and their effect upon matter. Typically, engineering mechanics is used to analyze and predict the acceleration and **deformation** (both elastic and plastic) of objects under known forces (also called loads) or stresses. Subdisciplines of mechanics include **statics**, **dynamics**, mechanics of materials, fluid mechanics, kinematics, and **continuum mechanics.**

Mechatronics and Robotics

10. Mechatronics is a combination of mechanics and electronics. It is an **interdisciplinary** branch of mechanical engineering, electrical engineering and software engineering that is concerned with integrating electrical and mechanical engineering to create **hybrid** systems. In this way, machines can be automated through the use of electric motors, servo-mechanisms, and other electrical systems in conjunction with special software. A common example of a mechatronics system is a CD-ROM drive. Mechanical systems open and close the drive, spin the CD and move the laser, while an optical system reads the data on the CD and converts it to bits. Integrated software controls the process and communicates the contents of the CD to the computer.

Structural Analysis

11. Structural analysis is the branch of mechanical engineering (and also civil engineering) devoted to examining why and how objects fail and to fixing the objects and their performance. Structural failures occur in two general modes: static failure, and fatigue failure. Static failure occurs when, upon being loaded (having a force applied) the object being analyzed either breaks or is deformed plastically, depending on the criterion for failure. Fatigue failure occurs when an object fails after a number of repeated loading and unloading cycles. Fatigue failure occurs because of imperfections in the object: a microscopic crack on the surface of the object, for instance, will grow slightly with each cycle (propagation) until the crack is large enough to cause ultimate failure.

(Source: https://en.wikipedia.org/wiki/Mechanical_engineering#Education)

New Words and Expressions

algorithm	*n.*	算法
biomechanics	*n.*	生物力学
biomechatronics	*n.*	生物机械电子学
bionanotechnology	*n.*	生物纳米技术
composite	*n.*	复合材料
continuum mechanics		连续介质力学（又作 mechanics of continuous media）
deformation	*n.*	变形
dynamics	*n.*	动力学
hybrid	*adj.*	混合的
interdisciplinary	*adj.*	跨学科的
iterate	*v.*	迭代
kinematics	*n.*	运动学
mating part		配件
mechatronics	*n.*	机械电子学
metallurgical	*adj.*	冶金（学）的
nanotechnology	*n.*	纳米技术
prototype	*n.*	原型
simulation	*n.*	模拟，仿真
statics	*n.*	静力学
thermodynamics	*n.*	热力学
viscoelasticity	*n.*	黏弹性

Comprehension Checks

Work in pairs and discuss the following questions. Then share your ideas with the rest of the class.

1. Could you briefly introduce mechanical engineering as a branch of science? (para.1)
2. What might be some potential advantages of incorporating CAE programs into the existing design and analysis processes? (para.4)
3. What benefits could CAE program bring along? (para. 6)
4. What's the working mechanism of CD-ROM drive? (para. 10)
5. Under what condition will fatigue failure occur and what might be the underlying causes? (para. 11)

Vocabulary Tasks

I. Complete the sentences below. The first letter of the missing word has been given.

1. Mechatronics is an i________ branch of mechanical engineering, electrical engineering and software engineering.
2. A technical drawing can be a computer model or hand-drawn s________ showing all the dimensions necessary to manufacture a part.
3. They also utilize sophisticated optimization a________ to more intelligently explore possible designs.
4. Engineering mechanics is used to analyze and predict the d________ of objects under known forces or stresses.
5. Engineering t________ is concerned with changing energy from one form to another.

II. Match each word with the correct definition.

1. prototype	a. the science of making or working with things that are so small that they can only be seen using a powerful microscope
2. deformation	b. in the form of a diagram that shows the main features or relationships but not the details
3. composite	c. the process or result of changing and spoiling the normal shape of something
4. schematic	d. the scientific study of the physical movement and structure of living creatures
5. kinematics	e. the first design of something which other forms are copied or developed
6. nanotechnology	f. the branch of mechanics concerned with motion without reference to force or mass
7. thermodynamics	g. an object or item which is made up of several different things, parts, or substances
8. biomechanics	h. the science that deals with the relations between heat and other forms of energy

III. Fill in the blanks with the words given below. Change the forms where necessary.

dynamics	deformation	algorithm	mechanics
interdisciplinary	schematic	pertinent	simulation

1. The scientist developed one model to ________ a full year of the globe's climate.
2. Most ________ devices require oil as a lubricant (润滑油).
3. In the operation of the curved box-girder, bending and torsion ________ has also been quite pronounced.
4. ________ information will be forwarded to the appropriate party.
5. Bioinformatics is the ________ of computational molecular biology and information processing science.
6. Electric products shall be attached with the ________ diagram and circuit diagram.
7. Unicode encoding and ________ are used for internal communications in the AIX system.
8. Scientists observe the same ________ in fluids.

Text B

Technical drawing

1. Technical drawing, drafting or drawing, is the act and discipline of composing drawings that visually communicate how something functions or is constructed.

2. Technical drawing is essential for communicating ideas in industry and engineering. To make the drawings easier to understand, people use familiar symbols, perspectives, units of measurement, **notation** systems, visual styles, and page **layout**. Together, such conventions constitute a visual language and help to ensure that the drawing is **unambiguous** and relatively easy to understand. Many of the symbols and principles of technical drawing are **codified** in an international standard called ISO 128.

3. The need for precise communication in the preparation of a functional document distinguishes technical drawing from the expressive drawing of the visual arts. Artistic drawings are subjectively interpreted; their meanings are multiply determined. Technical drawings are understood to have one intended meaning.

4. A drafter, drafts person, or draughtsman is a person who makes a drawing (technical or expressive). A professional drafter who makes technical drawings is sometimes called a drafting technician. Professional drafting is a desirable and necessary function in the design and manufacture of complex mechanical components and machines. Professional drafts person bridge the gap between engineers and manufacturers and contribute experience and technical **expertise** to the design process.

5. Methods of technical drawing includes manual or by **instrument**, and computer aided design. The basic drafting procedure is to place a piece of paper (or other material) on a smooth surface with right-angle corners and straight sides—typically a drawing board. A sliding **straightedge** known as a T-square is then placed on one of the sides, allowing it to be slid

across the side of the table, and over the surface of the paper.

6. "Parallel lines" can be drawn simply by moving the T-square and running a pencil or technical pen along the T-square's edge. The T-square is used to hold other devices such as set squares or triangles. In this case, the drafter places one or more triangles of known angles on the T-square—which is itself at right angles to the edge of the table—and can then draw lines at any chosen angle to others on the page. Modern drafting tables come equipped with a drafting machine that is supported on both sides of the table to slide over a large piece of paper. Because it is secured on both sides, lines drawn along the edge are guaranteed to be parallel.

7. In addition, the drafter uses several technical drawing tools to draw curves and circles. Primary among these are the compasses, used for drawing simple **arcs** and circles, and the French curve, for drawing curves. A **spline**① is a rubber coated **articulated** metal that can be manually bent to most curves.

8. Drafting **templates** assist the drafter with creating recurring objects in a drawing without having to reproduce the object **from scratch** every time. This is especially useful when using common symbols; i.e. in the context of stagecraft, a lighting designer will draw from the USITT standard library of lighting fixture symbols to indicate the position of a common fixture across multiple positions. Templates are sold commercially by a number of vendors, usually customized to a specific task, but it is also not uncommon for a drafter to create his own templates.

9. This basic drafting system requires an accurate table and constant attention to the positioning of the tools. A common error is to allow the triangles to push the top of the T-square down slightly, thereby throwing off all angles. Even tasks as simple as drawing two angled lines meeting at a point require a number of moves of the T-square and triangles, and in general, drafting can be a time-consuming process.

10. A solution to these problems was the introduction of the mechanical "drafting machine", an application of the **pantograph** (sometimes referred to incorrectly as a "pentagraph" in these situations) which allowed the drafter to have an accurate right angle at any point on the page quite quickly. These machines often included the ability to change the angle, thereby removing the need for the triangles as well.

11. In addition to the mastery of the mechanics of drawing lines, arcs and circles (and text) onto a piece of paper—with respect to the detailing of physical objects—the drafting effort requires a thorough understanding of **geometry**, **trigonometry**② and **spatial** comprehension, and in all cases demands precision and accuracy, and attention to detail of high order.

---

① 样条，绘制曲线的一种绘图工具，是富有弹性的细长条。

② 三角学，研究平面三角形和球面三角形边角关系的数学学科。

12. Today, the mechanics of the drafting task have largely been automated and accelerated through the use of computer-aided design systems (CAD).

13. There are two types of computer-aided design systems used for the production of technical drawings: two dimensions ("2D") and three dimensions ("3D").

14. 2D CAD systems such as AutoCAD or MicroStation replace the paper drawing discipline. The lines, circles, arcs, and curves are created within the software. It is down to the technical drawing skill of the user to produce the drawing. There is still much scope for error in the drawing when producing first and third angle **orthographic projections**①, **auxiliary** projections and cross sections②. A 2D CAD system is merely an electronic drawing board. Its greatest strength over direct to paper technical drawing is in the making of revisions. Whereas in a conventional hand drawn technical drawing, if a mistake is found, or a modification is required, a new drawing must be made from scratch, the 2D CAD system allows a copy of the original to be modified, saving considerable time. 2D CAD systems can be used to create plans for large projects such as buildings and aircraft but provide no way to check whether the various components will fit together.

15. A 3D CAD system (such as KeyCreator, Autodesk Inventor, or SolidWorks) first produces the geometry of the part; the technical drawing comes from user defined views of that geometry. Any orthographic, projected or sectioned view is created by the software. There is no scope for error in the production of these views. The main scope for error comes in setting the parameter of first or third angle projection and displaying the relevant symbol on the technical drawing. 3D CAD allows individual parts to be assembled together to represent the final product. Buildings, aircraft, ships, and cars are modeled, assembled, and checked in 3D before technical drawings are released for manufacture.

(Source:
https://en.wikipedia.org/wiki/Technical_drawing#Applications_for_technical_drawing)

New Words and Expressions

arc	*n.*	弧，弧线
articulated	*adj.*	铰接式的
auxiliary	*adj.*	辅助的
codify	*v.*	编纂
expertise	*n.*	专业知识

① 正射投影：由一点放射的投射线所产生的投影称为中心投影，由相互平行的投射线所产生的投影称为平行投影，平行投射线垂直于投影面的称为正射投影。

② 横断面，是指从垂直于物体的轴心线的方向切断物体后所露出的表面。

from scratch		从头做起
instrument	*n.*	工具，仪器
geometry	*n.*	几何学
layout	*n.*	布置图，规划图，图面配置
notation	*n.*	符号，记号
orthographic projection		正射投影
pantograph	*n.*	缩放仪，比例绘图仪
spatial	*adj.*	空间的
spline	*n.*	样条
straightedge	*n.*	平尺
template	*n.*	模板
trigonometry	*n.*	三角学
unambiguous	*adj.*	明确的

Comprehension Questions

1. The major difference between expressive drawing and technical drawing is ________.
   A. whether the drawing is subjectively interpreted or objectively interpreted
   B. whether the drawing is expensive or not
   C. whether the drawing has only one or more than one meaning
2. What does the underlined word "manually" (line 3, para. 7) probably mean?
   A. 手动地　　B. 轻易地　　C. 自动地
3. Which of the following statement is FALSE about templates?
   A. Many vendors sell templates to gain economic benefits.
   B. With drafting templates, drafters have to reproduce the object from the beginning every time.
   C. Drafters often create templates by themselves.
4. Compared with conventional hand drawn technical drawing, 2D CAD system wins over as ________.
   A. it is no more than a drawing board
   B. it can save much time in making revisions
   C. it can check whether the various components can fit together
5. What's the typical feature of 3D CAD system?
   A. The main scope for error comes in producing first and third angle orthographic projections.
   B. After released for manufacture, technical drawings are checked in 3D.
   C. 3D CAD allows individual parts to be assembled together to represent the final product.

Grammar Coach

Inversion

英语的基本语序是“主语+谓语+其他成分”，但有时出于强调或修辞的需要，会将主谓倒置，即将整个谓语或者谓语的一部分放在主语的前面构成倒装句（inversion）。科技英语文本为了强调某一概念和信息，或为了使句子保持平衡，会较多地使用倒装句。

倒装包括完全倒装（full inversion）和部分倒装（partial inversion）。完全倒装是将整个谓语部分置于主语之前，其他成分位置不变；部分倒装是将谓语中的助动词或情态动词置于主语之前，其他成分位置不变。

以下将通过举例对科技英语文本中采用倒装句的情况进行说明，如表 8–1 所示。

表 8–1　常见倒装句型及例句

	常见句式说明	例　句
完全倒装	here、there、up、down、in、out、off、away 等词开头的句子，表示强调	**There** exists a very important relationship between current, voltage and resistance.
	表示地点的介词短语作状语位于句首，强调地点	**Up** went the rocket, carrying the satellite on top of it.
	表语置于句首，起强调作用，同时也可以保持句子平衡	**Present at the meeting** were 500 scientists.
部分倒装	表示否定、唯一概念的副词和连词置于句首构成部分倒装	**No longer** should we waste the non-renewable resources because they are very valuable. **Only** when x>5 does the inequality holds.
	含有 had、were、should 的虚拟条件句，省略 if，将 had、were、should 置于动词之前，强调假设的条件	**Should** you observe carefully, you would discover the difference between the two phenomena.
	当 so、neither、nor 引导从句且位于句首时，将从句中与主句重复的谓语的一部分置于主语之前，构成部分倒装，强调说明从句的主语和主句的主语具有相同的特点	The light from the sun is incoherent light, **so** is the light from an incandescent bulb.
	as 引导的让步状语从句	A giant industrial country **as** China is, it still faces the risk of de-industrialization.

Exercise

Translate the following sentences into English

1. 只有节约资源，我们才能促进环境的可持续发展。

2. 下面是几个和变压器（transformers）有关的例子。
3. 直到 19 世纪，研究人员才得出结论：只有两种电。
4. 一些天体（celestial body）绕太阳运行。
5. 石油资源是很宝贵的，煤炭资源也很宝贵。
6. 这个表格列举了我国的几种稀缺能源。
7. 如果有更多的清洁可再生能源，我们就会减少对一次性能源的依赖。

Section C

Power Engineering

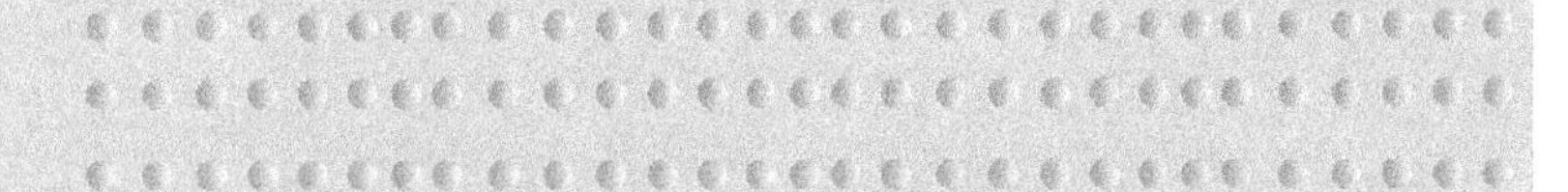

Unit 9

Heat Pumps and Heat Pipes

Text A

Heat Pump Systems

1. For climates with **moderate** heating and cooling needs, heat pumps offer an energy-efficient alternative to **furnaces** and air conditioners. Like your refrigerator, heat pumps use electricity to move heat from a cool space to a warm space, making the cool space cooler and the warm space warmer. During the heating season, heat pumps move heat from the cool outdoors into your warm house and during the cooling season, heat pumps move heat from your cool house into the warm outdoors. Because they move heat rather than generate heat, heat pumps can provide equivalent space conditioning at as little as one quarter of the cost of operating conventional heating or cooling **appliances**.

2. There are three types of heat pumps: air-to-air, water source, and geothermal. They collect heat from the air, water, or ground outside your home and concentrate it for use inside.

3. The most common type of heat pump is the air-source heat pump, which transfers heat between your house and the outside air. Today's heat pump can reduce your electricity use for heating by approximately 50% compared to electric resistance heating such as furnaces and baseboard heaters. High-efficiency heat pumps also **dehumidify** better than standard central air conditioners, resulting in less energy usage and more cooling comfort in summer months. Air-source heat pumps have been used for many years in nearly all parts of the United States, but until recently they have not been used in areas that experienced extended periods of **subfreezing** temperatures. However, in recent years, air-source heat pump technology has advanced so that it now offers a **legitimate** space heating alternative in colder regions.

4. For homes without ducts, air-source heat pumps are also available in a ductless version called a mini-split heat pump. In addition, a special type of air-source heat pump called a "reverse cycle chiller" generates hot and cold water rather than air, allowing it to be used with **radiant** floor heating systems in heating mode.

5. Geothermal (ground-source or water-source) heat pumps achieve higher efficiencies by transferring heat between your house and the ground or a nearby water source. Although they cost more to install, geothermal heat pumps have low operating costs because they take advantage of relatively constant ground or water temperatures. Geothermal heat pumps have some major advantages. They can reduce energy use by 30%-60%, control humidity, are **sturdy** and reliable, and fit in a wide variety of homes. Whether a geothermal heat pump is appropriate for you will depend on the size of your lot, the **subsoil**, and the landscape. Ground-source or water-source heat pumps can be used in more extreme climates than air-source heat pumps, and customer satisfaction with the systems is very high.

6. A new type of heat pump for residential systems is the absorption heat pump, also called a gas-fired heat pump. Absorption heat pumps use heat as their energy source, and can be driven with a wide variety of heat sources.

7. A number of innovations are improving the performance of heat pumps.

8. Unlike standard **compressors**[①] that can only operate at full capacity, two-speed compressors allow heat pumps to operate close to the heating or cooling capacity needed at any particular moment. This saves large amounts of electrical energy and reduces compressor wear. Two-speed heat pumps also work well with zone control systems. Zone control systems, often found in larger homes, use automatic **dampers** to allow the heat pump to keep different rooms at different temperatures.

9. Some models of heat pumps are equipped with variable-speed or dual-speed motors on their indoor fans (blowers), outdoor fans, or both. The variable-speed controls for these fans attempt to keep the air moving at a comfortable velocity, minimizing cool drafts and maximizing electrical savings. It also minimizes the noise from the blower running at full speed.

10. Many high-efficiency heat pumps are equipped with a **desuperheater**[②], which recovers waste heat from the heat pump's cooling mode and uses it to heat water. A desuperheater-equipped heat pump can heat water 2 to 3 times more efficiently than an ordinary electric water heater.

11. Another advance in heat pump technology is the **scroll** compressor[③], which consists

① 压缩机，指将低压气体提升为高压气体的一种从动的流体机械，是制冷系统的心脏。

② 减温器，指用水作冷却介质调节过热式再热汽温的装置，其作用是控制和保持过热汽温或再热汽温为规定值，并防止过热器、再热器管壁受热。

③涡旋式压缩机，指由一个固定的渐开线涡旋盘和一个呈偏心回旋平动的渐开线运动涡旋盘组成的可压缩容积的压缩机。

of two **spiral**-shaped scrolls. One remains stationary, while the other orbits around it, compressing the refrigerant by forcing it into increasingly smaller areas. Compared to the typical piston compressors, scroll compressors have a longer operating life and are quieter. According to some reports, heat pumps with scroll compressors provide 10° to 15°F (5.6° to 8.3°C) warmer air when in the heating mode, compared to existing heat pumps with piston compressors.

(Source: https://energy.gov/energysaver/heat-pump-systems)

New Words and Expressions

appliance	*n.*	器械，装置
compressor	*n.*	压气机，压缩机
damper	*n.*	（调节空气，控制炉火燃烧的）风门，气闸
desuperheater	*n.*	减温器
dehumidify	*v.*	除湿，使干燥
furnace	*n.*	熔炉，火炉
legitimate	*adj.*	合理的，合法的
moderate	*adj.*	适度的，中等的
radiant	*adj.*	辐射的
scroll	*n.*	涡卷形，涡带
spiral	*adj.*	螺旋形的
sturdy	*adj.*	坚固的，耐用的
subfreezing	*adj.*	低于冰点的
subsoil	*n.*	（土壤的）底土，心土

Comprehension Checks

Work in pairs and discuss the following questions. Then share your ideas with the rest of the class.

1. What's the advantage of heat pumps over conventional heating or cooling appliances? (para.1)
2. Why "reverse cycle chiller" can be used with radiant floor heating systems? (para. 4)
3. What should be taken into consideration in choosing a geothermal heat pump? (para. 5)
4. What's the difference between standard compressors and two-speed compressors? (para. 8)
5. Could you briefly introduce the working mechanism of scroll compressors? (para. 11)

Vocabulary Tasks

I. Fill in the blanks with the words given below. Change the forms where necessary.

moderate	compressor	geothermal	supplement	incorporate
radiant	combustion	innovate	appliance	velocity

1. The gasoline vaporizes and mixes with the air to form a highly ________ mixture.
2. Most drugs offer either no real improvement or, at best, only ________ improvements.
3. A ________ is a machine or part of a machine that squeezes gas or air and makes it take up less space.
4. This technical ________ is essential to exploiting new energy.
5. Continuous energy flows include ________, solar, tidal and wind energy.
6. The water ________ of the flood stood at 3.6 meters per second.
7. In addition to the light that's visible to us, the sun also ________ ultraviolet and infrared light.
8. In recent years, solar energy and wind power have been a nice energy ________ while the sun is shining and the wind is blowing.
9. There are few heat pump manufacturers that ________ both types of heat supply in one box.
10. He could also learn to use the vacuum cleaner, the washing machine and other household ________.

II. Match each word with the correct definition.

1. appliance	a. the speed at which something moves in a particular direction
2. compressor	b. colder than the temperature at which water freezes
3. geothermal	c. a device or machine in your home that you use to do a job such as cleaning or cooking
4. velocity	d. sent out in rays from a central point
5. subfreezing	e. strong and not easily damaged
6. stationary	f. a machine or part of a machine that squeezes gas or air and makes it take up less space
7. radiant	g. connected with the natural heat or rock deep in the ground
8. sturdy	h. not moving or not intended to move

III. Below is a commercial of heat pump. Put the sentences in the right order.

A SMART ALTERNATIVE TO AN AIR CONDITIONER

(   ) Heat pumps look and function the same as air conditioners for cooling, but in cool months when heat is called for, they reverse operation to provide warmth for your home.

(   ) Heating, cooling and dehumidifying—heat pumps have it all.

(   ) If you live in a colder environment, electric-powered heat pumps are great in combination with your oil- or gas-fueled furnace as a Hybrid Heat solution, which can result in significant savings on your overall heating costs.

(   ) Except, perhaps, a name that does their versatility justice.

ext B

Heat Pipes

1. Heat pipes (HPs)① are one of the most efficient ways to move heat, or thermal energy, from one point to another. These two-phase systems are typically used to cool areas or materials, even in outer space. Heat pipes were first developed for use by Los Alamos National Laboratory② to supply heat to and remove waste heat from energy conversion systems.

2. Today, heat pipes are used in a variety of applications from space to handheld devices that fit in your pocket. Heat pipes are present in the cooling and heat transfer systems found in computers, cell phones, and satellite systems.

3. A heat pipe is a simple tool, but how it works is quite **ingenious**.

4. These devices are sealed vessels that are **evacuated** and backfilled with a working fluid, typically in a small amount. The pipes use a combination of evaporation and condensation of this working fluid to transfer heat in an extremely efficient way.

5. The most common HP is **cylindrical** in cross-section, with a **wick** on the inner **diameter**. Cool working fluid moves through the wick from the colder side (condenser③) to the hotter side (evaporator④) where it vaporizes. This vapor then moves to the condenser's heat

---

① 热导管，或称热管，是一种具有快速均温特性的特殊材料，具有优异的热超导性能。

② 洛斯阿拉莫斯国家实验室，英文缩写为 LANL，简称为阿拉莫斯实验室，以前因为保密原因对外称为 Y 地点，隶属于美国能源部（DOE）。该实验室最初由加州大学全权负责运行管理，后于 2007 年变更为由洛斯阿拉莫斯国家安全机构（由加州大学等机构共同构成）主管。

③ 冷凝器，制冷系统的机件，属于换热器的一种，能把气体或蒸气转变成液体，将管子中的热量以很快的方式传到管子附近的空气中。冷凝器的工作过程是个放热的过程，所以冷凝器温度都是较高的。

④ 蒸发器，制冷四大件中很重要的一个部件，低温的冷凝液体通过蒸发器与外界的空气进行热交换，气化吸热，达到制冷的效果。

sink, bringing thermal energy along with it. The working fluid condenses, releasing its latent heat in the condenser, and then repeats the cycle to continuously remove heat from part of the system.

6. The temperature drop in the system is minimal due to the very high heat transfer **coefficients** for boiling and condensation. Effective thermal conductivities can approach 10,000 to 100,000 W/(m • k) for long HPs, in comparison with roughly 400 W/(m • k) for copper. The choice of material varies depending on the application, and has led to pairings such as **potassium** with **stainless steel**, water with copper, and ammonia with **aluminum**, steel and **nickel**. Benefits include passive operation and very long life with little to no maintenance.

How a Heat Pipe Works

7. A heat pipe consists of a working fluid, a wick structure, and a vacuum-tight **containment** unit (envelope). As discussed earlier, the heat input vaporizes the working fluid in liquid form at the wick surface in the evaporator section.

8. Vapor and its associated latent heat flow toward the colder condenser section, where it condenses, giving up the latent heat. **Capillary** action then moves the condensed liquid back to the evaporator through the wick structure. Essentially, this operates in the same way as how a sponge soaks up water.

9. Phase-change processes and the two-phase flow circulation in the HP will continue as long as there is a large enough temperature difference between the evaporator and condenser sections. The fluid stops moving if the overall temperature is uniform, but starts back up again as soon as a temperature difference exists. No power source (other than heat) is needed.

10. In some cases, when the heated section is below the cooled section, gravity is used to return the liquid to the evaporator. However, a wick is required when the evaporator is above the condenser on earth. A wick is also used for liquid return if there is no gravity, such as in NASA's micro-gravity applications.

What are the Benefits of a Heat Pipe?

11. There are some universal benefits of how a heat pipe works across almost all applications. Heat pipes have high effective thermal conductivity. They can transfer heat over long distances, with minimal temperature drop. Also, heat pipe has long life with no maintenance. The vacuum seal prevents liquid losses, and protective coatings can give each device a long-lasting protection against corrosion. What's more, heat pipes have lower costs. By lowering the operating temperature, these devices can increase the Mean Time Between Failure (MTBF) for electronic assemblies. In turn, this lowers the maintenance required, and the replacement costs. In HVAC systems, they can reduce the energy required for heating and

air conditioning, with payback times of a couple of years.

(Source: https://www.1-act.com/innovations/heat-pipes/)

New Words and Expressions

aluminum	*n.*	铝
capillary	*n.*	毛细血管
coefficient	*n.*	系数
containment	*n.*	容量，包容性
cylindrical	*adj.*	圆柱形的，圆柱体的
diameter	*n.*	直径
evacuate	*v.*	除清，抽空
ingenious	*adj.*	精巧的，巧妙的
nickel	*n.*	镍
potassium	*n.*	钾
stainless steel		不锈钢
wick	*n.*	灯芯

Comprehension Questions

1. Which of the following statement is FALSE about heat pipes?
   A. Heat pipes are simple tools and the devices are sealed containers.
   B. Heat pipes are filled with a large amount of working fluid.
   C. Heat pipes transfer in an efficient way by combining evaporation and condensation of the working fluid.
2. The passage mentions a sponge soaking water in order to explain ________.
   A. the working process of the heat pipe
   B. the structure of the heat pipe
   C. the feature of the heat pipe
3. What does the underlined word "uniform" (line 3, para. 9) probably mean?
   A. 不一致的　　B. 平均的　　C. 一致的
4. Wicks become unnecessary when ________.
   A. the evaporator is above the condenser on earth
   B. the heated section is below the cooled section
   C. there is no gravity for liquid return
5. Which of the following best states the benefit of heat pipes?
   A. Their thermal conductivity effect is low.

B. They don't need maintenance to keep a long life.

C. They cost higher than other pipes.

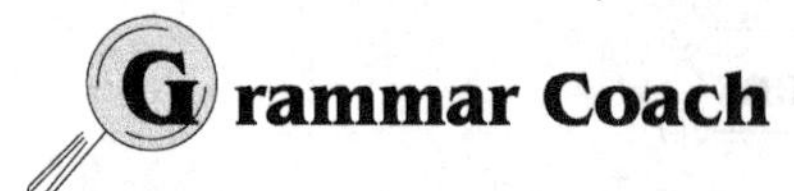

IT-Pattern

IT-Pattern，即 IT 结构，主要指 it 在句中作形式主语、形式宾语，以及无指代的情况（it 引导的强调句型）。在科技英语文章中，经常会出现大量 IT 结构，因为 IT 结构既可以平衡句子结构，避免因主语或宾语过长而造成句子头重脚轻的现象，又能够很好地强调句子重点，符合科技英语准确、客观、简明的写作特点。

1. it 作形式主语

it 作形式主语，即将 it 置于句首作形式主语，把真正的主语（不定式，that 从句或者动名词）置于句末。在科技英语中，it 作形式主语主要分为以下 3 种句型：

（1）It+系动词+adj./done/n.+to do/ doing/that…

例如： It is evident that people are threatened by the lack of energy resources.

It is worthwhile adhering to the principle of sustainable development.

It is a simple matter to take advantage of geothermal energy.

（2）It + 系动词 + adj/n.+of/for sb. +to do sth.

例如： It is important for the energy industry to increase source utilization rate.

It is foolish of us to exploit oil without restraint.

（3）It's no use/good +doing sth.

例如： It's no good excessively depending on domestic energy resources.

It's no use only recognizing the potential of wind energy but taking no actions.

2. it 作形式宾语

it 作形式宾语，即当不定式、动名词或从句在句子中作宾语时，将 it 置于宾语的位置，而将真正的宾语放在句尾。从而保持句子结构平衡，避免句式结构的混乱。在科技英语中，it 作形式宾语主要分为以下 2 种句型:

(1) 主语+think/find/consider/make/feel/believe it + adj./n. + doing/ to do/that…

例如： Machines make it possible that people do more work with less energy.

We believe it practical to utilize tidal energy for its low cost.

(2) 主语+depend(rely) on it/see to it/take it for granted +that…

例如： Human beings should see to it that they leave a clean environment for later generations.

Don't take it for granted that natural resources are inexhaustible.

3. it 引导的强调句型

it 引导的强调句型，即“It is/was +被强调部分+that/who/whom...”，将被强调的部分放在前面，其他部分置于 that/who/whom 后面。被强调部分可以是主语、宾语、表语或状语。

例如： It is to obtain on space that many satellites have been launched.
It may be this flaw that leads to the failure of that computer.

Exercise

Please use IT-Pattern to translate the following sentences into English.

1. 很明显，氧气是一种无色无味的气体。
2. 已经证实的是，能量只能转化为其他形式。
3. 高价能源这一趋势似乎会持续下去。
4. 提高能源利用率是有必要的。
5. 大肆破坏环境是危险的。
6. 破坏生物多样性是不好的。
7. 风力发电站的建设使风能的利用成为可能。
8. 矿物学家发现勘探煤炭资源十分重要。
9. 在 1946 年，人类发明了世界上第一台电子计算机。
10. 18 世纪英国发明的蒸汽动力使工业革命成为可能。

Unit 10

Automation Control

Text A

What is Building Automation

The Basics

1. Complete autonomous control of an entire **facility** is the goal that any modern automation system attempts to achieve. The distributed control system is the computer networking of electronic devices designed to monitor and control the mechanical, security, fire, lighting, HVAC① and humidity control and **ventilation** systems in a building or across several campuses.

2. The core function of the Building Automation System (BAS) is to keep building climate within a specified range, light rooms based on an occupancy schedule, monitor performance and device failures in all systems and provide malfunction alarms. Automation systems reduce building energy and maintenance costs compared to a non-controlled building. Typically they are financed through energy and insurance savings and other savings associated with pre-emptive maintenance and quick detection of issues.

3. A building controlled by a BAS is often referred to as an intelligent building or "smart building". Commercial and industrial buildings have historically relied on **robust** proven **protocols** like BACnet②.

4. Almost all multi-story green buildings are designed to accommodate a BAS for the

---

① HVAC，heating, ventilation and air conditioning 的英文缩写，译为供热、通风与空气调节。

② BACnet，用于智能建筑的通信协议，是国际标准化组织（ISO）、美国国家标准学会（ANSI）及美国采暖、制冷与空调工程师学会（ASHRAE）定义的通信协议。

energy, air and water **conservation** characteristics. Electrical device demand response is a typical function of a BAS, as is the more sophisticated ventilation and humidity monitoring required of "tight" **insulated** buildings. Most green buildings also use as many low-power DC devices as possible, typically integrated with power over Ethernet wiring, so by definition always accessible to a BAS through the Ethernet connectivity. Even a Passivhaus design intended to consume no net energy whatsoever will typically require a BAS to manage heat capture, shading and **venting**, and scheduling device use. Figure 10–1 shows an example of an integrated building automation system.

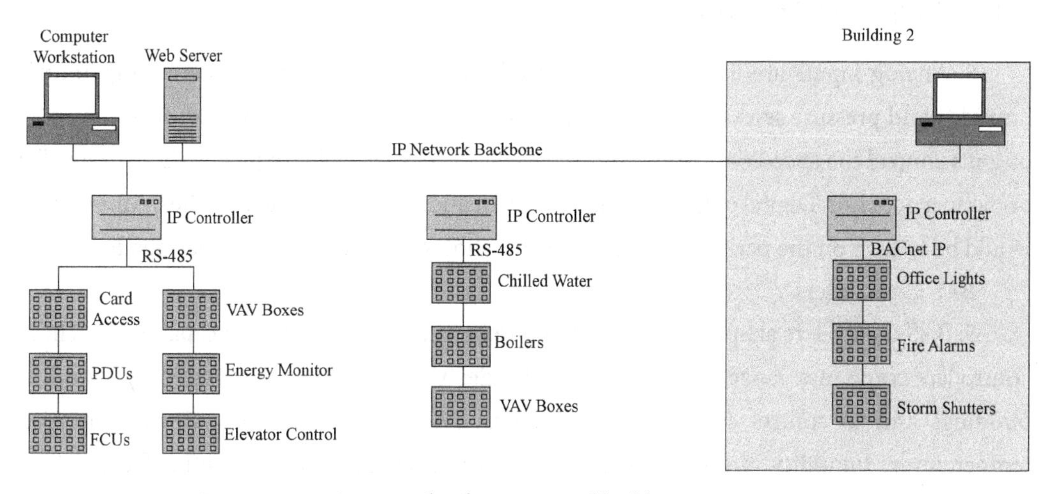

Figure 10–1 An example of an integrated building automation system

Buses and Protocols

5. Most building automation networks consist of a primary and secondary bus which connect high-level controllers with lower-level controllers, input/output devices and a user interface devices. ASHRAE's① open protocol BACnet or the open protocol LonTalk specifies how most such devices interoperate. Modern systems use SNMP② to track events, building on decades of history with SNMP-based protocols in the computer networking world.

6. Physical connectivity between devices was historically provided by dedicated optical fiber③, Ethernet④, ARCNET, RS-232, RS-485 or a low-**bandwidth** special purpose wireless network. Modern systems rely on standards-based multi-protocol **heterogeneous** networking. These accommodate typically only IP-based networking but can make use of any existing

① ASHRAE，成立于 1894 年，在全球拥有超过 54 000 名成员。该学会及其成员专注于建筑系统、能源效率、室内空气质量、制冷和行业内的可持续性。

② SNMP，简单网络管理协议，由一组网络管理的标准组成，包含一个应用层协议、数据库模型和一组资料物件。

③ 光纤，光导纤维的简称，是一种由玻璃或塑料制成的纤维，可作为光传导工具。

④ 以太网，由 Xerox 公司创建并由 Xerox、Intel 和 DEC 公司联合开发的基带局域网规范，是当今现有局域网采用的最通用的通信协议标准。

wiring, and also integrate power line networking over AC[①] circuits, power over Ethernet low power DC circuits, high-bandwidth wireless networks such as LTE and IEEE 802.11n and IEEE 802.11ac and often integrate these using the building-specific wireless mesh open standards.

7. Current systems provide **interoperability** at the application level, allowing users to mix-and-match devices from different manufacturers, and to provide integration with other **compatible** building control systems. These typically rely on SNMP, long used for this same purpose to integrate diverse computer networking devices into one coherent network.

Types of Inputs and Outputs

8. Analog inputs are used to read a variable measurement. Examples are temperature, humidity and pressure sensors. A digital input indicates if a device is turned on or not. Analog outputs control the speed or position of a device, such as a variable frequency drive or a valve or damper actuator. Digital outputs are used to open and close relays and switches. An example would be to turn on the parking lot lights when a photocell indicates it is dark outside.

BAS Controllers

9. BAS controllers are purpose-built computers with input and output capabilities. These controllers come in a range of sizes and capabilities to control devices commonly found in buildings and to control sub-networks of controllers. Inputs allow a controller to read temperatures, humidity, pressure, current flow, air flow, and other essential factors. The outputs allow the controller to send command and control signals to slave devices, and to other parts of the system. Inputs and outputs can be either digital or analog. Digital outputs are also sometimes called **discrete** depending on manufacturer.

10. Controllers used for building automation can be grouped in 3 categories: Programmable Logic Controllers (PLCs), System/Network Controllers, and Terminal Unit Controllers. However an additional device can also exist in order to integrate 3rd party systems (i.e., a stand-alone AC system) into a central building automation system).

(Source: http://www.controlservices.com/learning_automation.htm)

New Words and Expressions

automation	*n.*	自动化（技术），自动操作
bandwidth	*n.*	带宽
compatible	*adj.*	兼容的，相容的
conservation	*n.*	保护，保存
discrete	*adj.*	分离的，不相关联的

① AC，alternating current 的缩写，指大小和方向都发生周期性变化的电流，目前被广泛用于电力传输。

facility	*n.*	设备
heterogeneous	*adj.*	各种各样的
interoperability	*n.*	互操作性
insulate	*v.*	使隔离，使绝缘
protocol	*n.*	协议
robust	*adj.*	稳健的
vent	*n.*	排气口，通风口
ventilation	*n.*	通风

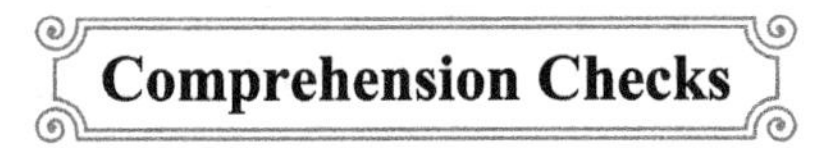

Comprehension Checks

Work in pairs and discuss the following questions. Then share your ideas with the rest of the class.

1. What's the core function of the BAS? (para.2)
2. Multi-story green buildings have common features. Could you explain some of them? (para.4)
3. Compared with older system designs, what would be the major advantages of modern automation system? (para.6)
4. Could you examplify each type of inputs and outputs? (para.8)
5. Could you briefly summarize the working mechanism of BAS controllers? (para.9)

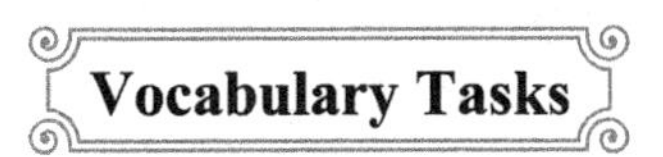

Vocabulary Tasks

I. Fill in the blanks with the words given below. Change the forms where necessary.

protocol	integrate	optimize	data	application
ventilation	autonomous	bandwidth	conserve	humidity

1. In the past decade, ________ has reduced the work force here by half.
2. The purpose of the ________ system is to supply and exhaust air for passenger safety and comfort.
3. Cheaper energy ________ techniques have been put into operation in the developed world.
4. This allows the ________ of two or more enterprise architectures for collaboration within the ecosystem.
5. To achieve sustainable development, the energy structure should be ________.
6. Ambient air temperature and ________ may also have an effect on the flammable limits.

7. High-signal, low-noise video reduces network ________ and storage requirements.
8. The invention would have a wide range of ________ in industry.
9. The new ________ technology would allow a rapid expansion in the number of TV channels.
10. Clean Development Mechanism can promote the developed countries to fulfill their commitments in Tokyo ________.

II. Match each word with the correct definition.

1. automation	a. the amount of water in the air
2. bandwidth	b. a device that regulates a machine or part of a machine
3. protocol	c. a band of frequencies used for sending electronic signals
4. ventilation	d. consisting of elements that are not of the same kind or nature
5. controller	e. modify to achieve maximum efficiency in storage capacity or time or cost
6. optimize	f. the act of supplying fresh air and getting rid of foul air
7. humidity	g. the first or original version of an agreement especially a treaty between countries
8. heterogeneous	h. the act of implementing the control of equipment with advanced technology

III. Below is an introduction to automation. Put the sentences in the right order.

(　　) It reduces the human factors, mental and physical, in production, and is designed to make possible the manufacture of more goods with fewer workers.

(　　) The development of automation in American industry has been called the "Second Industrial Revolution".

(　　) In the main, labour has taken the view that resistance to technical change is unfruitful.

(　　) Since it is expected that vast industries will grow up around manufacturing, maintaining, and repairing automation equipment.

(　　) Automation refers to the introduction of electronic control and automatic operation of productive machinery.

(　　) Eventually, the result of automation may well be an increase in employment.

(　　) Labor's concern over automation arises from uncertainty about the effects on employment, and fears of major changes in jobs.

5 Trends for the Industrial Automation and Control Market

1. For the past couple of years the industrial **segment** has been "challenged". Much of this was driven by the decline in the oil & gas market which had a significant **ripple effect**. Concurrently, a strong dollar hurt exports for many OEMs[①] and the agriculture market was soft, further hurting that segments equipment market. Overall, the industrial decline significantly impacted distributors and manufacturers. And a key product category, industrial automation and control, was the most impacted.

2. But according to manufacturers and distributors in many parts of the country, the industrial market appears to be turning. Growth, versus last year, is coming back. Some is due to an increase in oil **rigs**; natural gas projects are being quoted; companies are reinvesting either due to the need to do preventative maintenance or to generate a better ROI[②] on their cash and hence are upgrading; some to improve productivity and then there are companies investing to reduce energy costs through lighting improvements and companies that are investing into plant connectivity and the IoT (even if on an experimental basis).

3. Bill Lydon, who has been involved in the industrial automation and control market for almost 45 years and is the principal of Applied Marketing Concepts and writes on the automation control industry. Last year Bill talked to a wide range of users, suppliers and industry consultants, while attending over 20 industry conferences and related events. He recently shared his insights into trends in the segment on LinkedIn.

Internet of Things Technology will Cut Automation Costs

4. The discussions surrounding Internet of Things (IoT) concepts, and the technology impacting the industry, are becoming more <u>intense</u>. The ongoing development of technology and products is driving the Internet of Things with connected innovations including higher-performance processors, sensors, analytic software, vision systems, cloud computing[③], new communications (wireless & protocols) and highly distributed system architectures, among many other products. This, logically, should lead to lower cost and higher performance industrial automation systems.

---

① OEM (original equipment manufacturer)，原始设备制造商，是指受托厂商按来样厂商之需求与授权，按照其特定的条件而生产，所有的设计图等都完全依照来样厂商的设计来进行制造加工。

② ROI(return on investment)，投资回报率，指通过投资而应返回的价值，即企业从一项投资活动中得到的经济回报。

③ 云计算，指基于互联网的相关服务的增加、使用和交付模式，通常涉及通过互联网来提供动态易扩展且经常是虚拟化的资源。

Leaner & Flatter Architecture will Come to Automation

5. As I have forecasted in past trends articles, the evolution to **streamlined** 2-3 layer automation systems is starting to occur, increasing performance and lowering future software maintenance costs. This is a significant **simplification** of the 5 layer system model (typically expressed in the 5 level Purdue model) that the automation industry has been centered on for years. This trend is accelerating, with computing being driven down into more capable controllers, intelligent devices & sensors, and driven up to plant level computers and cloud hosted applications.

Open Industrial Automation Architectures will Simplify Integration

6. 2016 saw the emergence of open industrial system architectures, with key themes centering on computing at the edge, multivendor interoperable open systems, and tight integration with business systems. Older industrial automation architectures are currently not completely open but have communities of gated ecosystems, which presents an obstacle for multi-vendor integration. This new "open" way of thinking, along with the application of advanced technology, is intended to create more responsive, efficient, and flexible manufacturing that tightly integrate ecosystems of customers, suppliers, manufacturers, and distribution logistics. Achieving these goals requires **frictionless** communications and interaction between **enterprise** systems, manufacturing field I/O (inputs/outputs) including sensors, **actuators**, analyzers, drives, vision, video, and robotics to achieve increased manufacturing performance and flexibility. Creating superior integrated industrial automation systems in this environment requires fully open source communications, contextual data definitions & frameworks, and application interchange standards. These are some of the prominent recent initiatives: Industry 4.0, Industry 4.0 for Process, The Open Process Automation Forum.

Vendors must Embrace Portable Applications

7. One of the most challenging issues facing the automation industry, is the lack of multivendor portability of applications and this must change. The drivers behind open automation architecture initiatives recognize that without open ecosystems, providing portable applications between vendor platforms, innovation is **stifled**. Non-traditional suppliers are already demonstrating and selling products that are programmed with IoT software.

Democratizing Analytics and Edge/Cloud Historians Boost Big Data

8. Driven by a wide range of Internet of Things implementations, outside of industrial automation, significant developments have brought users a new generation of cloud services and tools to create analytics. Previously, much of the software used to accomplish advanced process control (APC), optimization, and predictive analytics has been difficult to use and expensive, but this is changing. Driven by a wide range of Internet of Things applications, these cloud-based tools, with refined integrated design environments, provide platforms for

users and industry experts to create and deploy economical analytics. These platforms significantly lower the cost of implementation and help broaden the range of applications where analytics can be applied, very similar to how Excel spread sheets empowered users to use computers more effectively. In addition to improving efficiency and productivity, more analytics can substantially better inform decision-making in order to improve and refine manufacturing processes.

9. Data historians have proven their value in the process industries, and now in discrete applications as well, but the biggest limiting factors have been the initial costs and ongoing lifecycle software maintenance. The use of the commercial cloud services is providing an economical means for many companies to leverage historic data and apply analytics where appropriate. I have talked with a number of major users that are collecting plant data, using OPC, OPC UA, and other gateways, and are delivering the data to standard cloud applications from providers such as Amazon Web Services (AWS) and Microsoft Azure cloud services. They are finding this approach efficient and highly flexible.

(Source: http://www.electricaltrends.com/2017/03/10-trends-industrial-automation-and-control-market.html)

New Words and Expressions

actuator	*n.*	执行器
enterprise	*n.*	企（事）业单位
frictionless	*adj.*	无摩擦的
rig	*n.*	钻井架，钻塔
ripple effect		涟漪效应
segment	*n.*	部分
simplification	*n.*	简化
stifle	*v.*	扼杀
streamline	*v.*	使简单化

Comprehension Questions

1. Which of the following factors does not influence the industrial segment?
   A. The decrease in the oil & gas market.
   B. The hindrance of strong dollar on exports.
   C. The robust agricultural market.
2. What does the underlined word “intense” (line 2, para. 4) probably mean?
   A. 不平衡的　　B. 强烈的　　C. 不稳定的

3. Which of the following is TRUE about the 2-3 layer automation system?
   A. It is more complicated compared with 5 layer system.
   B. It is a trend to simplify the system and the trend is accelerating.
   C. It will make the future software maintenance costs higher.
4. The passage states Excel increasing the Internet utilization efficiency in order to explain ________.
   A. the convenience and benefits brought by cloud-based tools
   B. the rapid development of Excel spread sheets
   C. the improvement of manufacturing process
5. What does the passage mainly talk about?
   A. The prospect for industrial automation and control.
   B. The challenges the industrial automation and control faces.
   C. The development of industrial automation and control.

rammar Coach

Elliptical Sentence

省略句（elliptical sentence），即将句子中的一个或多个成分省去而保持原句意思不变的语法现象。使用省略句可以避免重复，突出关键信息，平衡句子结构，并使上下文紧密相连。为了准确传递信息或突出重点信息，同时体现科技英语文章的简洁性，科技英语中常见省略句的表达方式，即省略主从复合句或并列句中重复的句子成分，包括主语、谓语、宾语、表语等。

当主从复合句中从句的主语、谓语、表语等与主句相同时，可以省略从句中与主句相同的成分；以两个简单句组成的并列句为例，当并列句中的两个句子有相同的成分时，也可以省略其中一个句子中相应的成分，从而使表达更加简洁，句子结构更加平衡。

1. 省略主语

例如： Scientists are looking for geothermal hot spots in night time but <u>so far have found none</u>.

该并列句的后半句省略了和前半句重复的主语 scientists。将其补充完整后为：so far scientists/they have found none。

2. 省略谓语

例如： The earth attracts the moon and <u>the moon the earth</u>.

该并列句的后半句省略了和前半句重复的谓语动词 attracts，使上下文衔接更加紧密，既表达了月球和地球相互吸引的关系，又使句子结构更加平衡。将其补充完整后为：the moon attracts the earth。

3. 省略表语

例如： Even if the equation is not satisfied, the forces obtained on the assumption that it is are accurate within a few percent.

句中 that 引导的从句中，it is 后省略了和让步状语从句中相同的表语 satisfied。

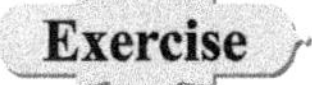

Translate the following sentences into English using elliptical sentences, and pay special attention to the parts in boldface.

1. **原因就是**有些能量在电线中失去。
2. 这个值大于**或等于 3**。
3. 虽然人们不能创造能量，但是能量能**从一种形式转变为另一种形式**。
4. 在特定条件下，热能可以**转化**为机械能，机械能也可以**转化**为热能。
5. **不论温度如何**，所有的物体都辐射电磁能（electromagnetic energy）。
6. 当两个温度不同的物体相接触时，热量从**较高温度的物体**传递到**较低温度的物体**。
7. 能量有不同的**形式**，但是这些**形式**可减少到两种主要状态。
8. 煤是天然物质，**但不属于矿物质**。

Unit 11

Engine Design

Text A

How Pistons Survive the Inferno inside Your Engine

1. The **slugs** of aluminium inside your engine live in a **fiery** hell. At full **throttle** and 6,000 rpm, a piston in a gasoline engine is subjected to nearly 10 tons of force every 0.02 second as repeated explosions heat the metal to more than 600 degrees **Fahrenheit**①.

2. These days, that cylindrical heads are hotter and more intense than ever, and it's only likely to get worse for pistons. As automakers chase higher efficiency, piston manufacturers are preparing for a future in which the most-**potent** naturally aspirated gasoline engines produce 175 horsepower per liter, up from 130 today. With **turbocharging** and increased outputs come even tougher conditions. In the past decade, piston operating temperatures have climbed 120 degrees, while peak cylinder pressures have **swollen** from 1,500 to 2,200 psi②.

3. A piston tells a story about the engine in which it resides. The crown can reveal the bore, the number of **valves**, and whether or not the fuel is directly injected into the cylinder. Yet a piston's design and technology can also say a lot about the wider trends and challenges facing the auto industry. To coin a maxim: as the automobile goes, so goes the engine; and as the engine goes, so goes the piston. In the quest for improved fuel economy and lower emissions, automakers are asking for lighter, lower-friction pistons with the **stamina** to endure tougher operating conditions. It is these three concerns—durability, friction, and mass—that consume the piston suppliers' workdays.

4. In many ways, gasoline-engine development is following the path laid out by **diesels** 15

---

① 华氏度，是用来计量温度的单位，符号为 F。

② psi: pounds per square inch，是压力计量单位。

years ago. To compensate for the 50-percent increase in peak cylinder pressures, some aluminium pistons now have an iron or steel insert to support the top ring. The hottest gasoline engines will soon require a cooling **gallery**, or an enclosed channel on the underside of the crown that's more efficient at removing heat than today's method of simply spraying the piston's underside with oil. The **squirters** shoot oil into a small opening on the bottom of the piston that feeds the gallery. The seemingly simple technology isn't easy to manufacture, though. Creating a hollow passage means casting the piston as two pieces and joining them via friction or **laser welding**.

5. Pistons account for at least 60 percent of the engine's friction, and improvements here have a direct impact on fuel consumption. Friction-reducing, **graphite-impregnated resin** patches screen-printed onto the skirt are now nearly universal. Piston supplier Federal-Mogul① is experimenting with a **tapered** face on the oil ring that allows a reduction in the ring tension without increasing oil consumption. Lower ring friction can unlock as much as 0.15 horsepower per cylinder.

6. Automakers are also hungry for new friction-reducing finishes between parts that rub or rotate against each other. The hard and slippery diamond-like coating, or DLC, holds promise for cylinder liners, piston rings, and wrist pins, where it can eliminate the need for bearings between the pin and connecting rod. But it's expensive and has few applications in today's cars.

7. "The manufacturers are often discussing DLC, but whether or not they will make it into production cars is a question mark," says Joachim Wagenblast, senior director of product development at Mahle②, a German auto-parts supplier.

8. Increasingly sophisticated computer modelling and more-precise manufacturing methods also enable more-complex shapes. In addition to the bowls, **domes**, and valve **indents** needed for clearance and to achieve a particular compression ratio, **asymmetric** skirts feature a smaller, **stiffer** area on the **thrust** side of the piston to reduce friction and stress concentrations. Flip a piston over and you'll see tapered walls scarcely more than 0.1 inch thick. Thinner walls require tighter control on tolerances that are already measured in **microns**, or thousandths of a millimeter.

9. Thinner walls also demand a better understanding of the thermal expansion of an object that sometimes has to warm from below freezing to several hundred degrees in a matter of seconds. The metal in your engine doesn't expand uniformly as it heats up, so optimizing tolerances requires the design experience and precise machining capabilities to create slight

① 辉门公司始创于 1899 年，公司总部位于美国密歇根州的绍斯菲尔德市，是一个全球性汽车零配件制造供应商。

② 德国马勒是众多国际汽车与发动机零部件制造商中的翘楚，为汽车与发动机行业提供高质量的零部件产品。众多革命性的创新使德国马勒成功地成为其客户的可靠伙伴和全球 30 大供应商之一。

eccentricities in the parts.

10. "Nothing we make is straight or round deliberately," says Keri Westbrooke, director of engineering and technology at Federal-Mogul. "We're always building in some compensation."

11. Diesel-engine pistons are undergoing their own evolution as peak cylinder pressures rise toward 3,600 psi. Mahle and Federal-Mogul are predicting a shift from **cast aluminium** to **forged steel** pistons. Steel is denser than aluminium but three times stronger, leading to a piston that is more resilient to higher pressures and temperatures with no increase in weight.

12. Steel enables a notable change in geometry by shortening the piston's compression height, defined as the distance from the center of the wrist pin to the top of the crown. This area accounts for 80 percent of the piston's weight, so shorter generally means lighter. Critically, a lower compression height doesn't just shrink the pistons. It also allows for a shorter and lighter engine block as the deck height is reduced.

(Source: http://www.popularmechanics.com/cars/g523/6-prototype-engines-to-get-your-brain-firing/)

New Words and Expressions

asymmetric	*adj.*	不对称的
cast aluminium		铸铝
diesel	*n.*	柴油机（diesel engine），内燃机车
dome	*n.*	圆屋顶形状，穹顶
eccentricity	*n.*	反常
Fahrenheit	*n.*	华氏温度，华氏度
fiery	*adj.*	火热的；激烈的；
forged steel		锻钢
gallery	*n.*	通道
graphite	*n.*	石墨，黑铅
impregnate	*v.*	灌注，使饱和
indent	*n/v.*	切割……使呈锯齿状
inferno	*n.*	地狱似的地方，恐怖景象，炽热
laser	*n.*	激光
micron	*n.*	微米
potent	*adj.*	有效的，有力的
resin	*n.*	树脂
slug	*n.*	（待加工的）金属小块，（形状不规则的铅制）子弹、弹丸

squirter *n.* 喷射器
stamina *n.* 耐力，持久力
stiff *adj.* 僵硬的
swollen （swell 的过去式和过去分词） 肿胀的
tapered *adj.* 锥形的
throttle *n.* 节流，节流门
thrust *n.* 猛推，推力
turbocharging *n.* 涡轮增压
valve *n.* 阀门，气门
welding *n.* 焊接

Comprehension Checks

Work in pairs and discuss the following questions. Then share your ideas with the rest of the class.

1. Why is the gasoline engine subjected to huge force? (para.1)
2. What would be the biggest concern of piston suppliers? (para.3)
3. What do you think of the prospect of DLC's entrance into car production?
4. What are some of the possible requirements of thinner walls? (para.8 & para.9)
5. What benefits can a piston bring if it bears a lower compression height? (para.12)

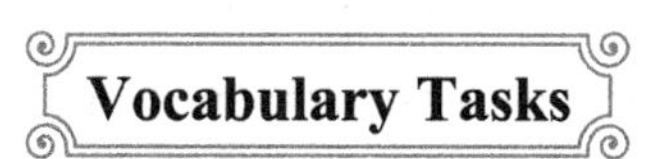

I. Fill in the blanks with the words given below. Change the forms where necessary.

resilient	potent	swell	diesel	application
piston	engine	sophisticated	stiff	thermal

1. The ________ is essentially a cylindrical plug that moves up and down in the engine cylinder.
2. Train doors have handles on the inside. They are ________ so that they cannot be opened accidentally.
3. Cotton is more resistant to being squashed and polyester is more ________.
4. Alcohol is a ________ drug that anaesthetizes（使麻醉）the brain.
5. Because the can is sealed, the gas builds up, causing the can to ________.
6. The ________ of a car or other vehicle is the part that produces the power which makes the vehicle move.

7. It's not difficult to discriminate a ________ engine from a gasoline engine.
8. As computer systems become even more ________, so too do the methods of those who exploit the technology.
9. Volcanic activity has created ________ springs and boiling mud pools.
10. The energy is released by combustion on the ________ of a match.

II. Match each word with the correct definition.

1. Fahrenheit	a. to push something suddenly or violently in a particular direction
2. laser	b. a scale for measuring temperature, in which water freezes at 32 degrees and boils at 212 degrees
3. piston	c. to make a substance spread through an area so that the area is filled with the substance
4. thrust	d. an unusual act or habit
5. impregnate	e. a cylinder or metal disc that is part of an engine, which causes various parts of the engine to move
6. eccentricity	f. a narrow beam of concentrated light produced by a special machine

III. Below is an introduction to heat piston. Put the sentences in the right order.

(　　) In a pump, the function is reversed and force is transferred from the crankshaft to the piston for the purpose of compressing or ejecting the fluid in the cylinder.

(　　) A piston is a component of reciprocating engines, reciprocating pumps, gas compressors and pneumatic cylinders, among other similar mechanisms.

(　　) In an engine, its purpose is to transfer force from expanding gas in the cylinder to the crankshaft via a piston rod and/or connecting rod.

(　　) In some engines, the piston also acts as a valve by covering and uncovering ports in the cylinder.

(　　) It is the moving component that is contained by a cylinder and is made gas-tight by piston rings.

Text B

Top 4 Improvements in Engine Design

1. Automotive engineers are constantly working on ways to improve the internal combustion engine and carry it into the future. How many other inventions do you know that have been continuously refined for more than 150 years?

2. In this article, we'll take a look at 4 of the biggest and most significant engine

improvements of all time. From forced **induction** to hybrid engines, we'll take a look at where engines have been, and hopefully get some insight on where they're headed.

Forced Induction

3. Benefits: More power without an increase in engine size

4. Drawbacks: Fuel consumption, turbo lag

5. An engine requires three things to generate motion: fuel, air, and ignition. **Cramming** more air into an engine will increase the power generated by the engine's pistons. A long-standing way to do that, and one that's becoming increasingly popular as of late, is to use forced induction. You may know this process better by the parts that do make it happen—**turbochargers**① and **superchargers**②.

6. In a forced induction engine, air is forced into the combustion chamber at a higher pressure than usual, creating a higher compression and more power from each stroke of the engine. Turbochargers and superchargers are essentially air compressors that **shove** more air into the engine.

7. Forced induction systems were used on aircraft engines long before they started being added to car engines in the 1920s. They are especially beneficial for small engines as they can generate a lot of extra power without increasing the engine's size or causing a dramatic drop in fuel economy.

8. A good example is the turbocharged Mini Cooper S, which only has a 1.6-liter engine but produces more than 200 horsepower in some applications. In addition, high-performance cars like the Porsche 911 Turbo or Corvette ZR-1 use forced induction to achieve tremendous gains in power.

9. And with fuel economy and emissions standards getting stricter, many car-makers are turning to forced induction on smaller engines instead of building larger engines. On the newest Hyundai Sonata, for example, the top engine one can buy is no longer a V6, but a turbo four-cylinder.

Fuel Injection

10. Benefits: Better throttle response, increased fuel efficiency, more power, easier starting

11. Drawbacks: More complexity and potentially expensive repairs

12. For decades, the preferred method for mixing fuel and air and depositing it into the

① 涡轮增压器实际上是一种空气压缩机，通过压缩空气来增加进气量。它是利用发动机排出的废气惯性冲力来推动涡轮室内的涡轮，涡轮又带动同轴的叶轮，叶轮压送由空气滤清器管道送来的空气，使之增压进入气缸。

② 增压器是活塞式航空发动机借以增加气缸进气压力的装置。进入发动机气缸前的空气先经过增压器压缩以提高空气的密度，使更多的空气充填到气缸里，从而增大发动机功率。

engine's combustion chamber was the carburettor[①]. Press the accelerator pedal to full throttle, and the carburettor allows more air and fuel into the engine.

13. Since the late 1980s, carburettors have been almost completely replaced by fuel injection, a far more sophisticated and effective system of mixing fuel and air. Fuel injectors spray gasoline into the air intake **manifold**, where fuel and air mix together into a fine mist. That mix is brought into the combustion chamber by valves on each cylinder during the intake process. The engine's on-board computer controls the fuel injection process.

14. So why did fuel injection replace the carburettor? To put it simply, fuel injection just works better in every aspect. Computer-controlled fuel-injected engines are easier to start, especially on cold days, when carburettors could make things tricky. Engines with fuel injection are also more efficient and more responsive to changes in the throttle.

15. They do have drawbacks in terms of their increased complexity. Fuel injection systems are more costly to repair than carburettors as well. However, they have become the industry standard for fuel delivery, and it doesn't look like carburettors which will be making a comeback anytime soon.

Direct Injection

16. Benefits: More power, better fuel economy

17. Drawbacks: More expensive to make, relatively new technology

18. Direct injection is a further refinement of the improvements made by fuel injection. As you may have guessed from its name, it allows fuel injection to "skip a step", which adds efficiency to the engine, and more power and improved fuel economy as a consequence.

19. On a direct injection engine, fuel is sprayed directly into the combustion chamber, not into the air intake manifold. Engine computers then make sure the fuel is burned exactly when and where it is needed, reducing waste. Direct injection provides a leaner mix of fuel, which burns more efficiently. In some ways it makes gasoline-powered engines more similar to diesel engines, which have always used a form of direct injection.

20. As we learned earlier, direct injection engines boast an increase in power and fuel economy over standard fuel injection systems. But they have their drawbacks as well. For one, the technology is a relatively new one, having come to market only in the last decade or so. More and more companies are starting to increase their use of direct injection, but it has yet to become the standard.

21. Sometimes, direct injection engines can exhibit the build-up of carbon deposits on the intake valves, which could cause reliability issues. Some car tuners have expressed difficulty

① 化油器，亦称为“汽化器”，发动机中用以使燃料与空气形成可燃混合物的部件。简单化油器主要由浮子室、量孔、喷管、喉管、节气门（俗称油门）等组成，喉管以上部分称为进气室，喉管以下部分称为混合室。

with modifying direct injection engines as well. Despite these issues, direct injection is the hot new technology in the automotive world right now. Expect to see it used in more and more cars as time goes on.

Hybrid Engines

22. Benefits: Fuel economy

23. Drawbacks: Higher initial cost, complexity

24. A combination of high gas prices, an increased awareness of the environment among drivers, and government regulations raising fuel economy and emissions standards have forced engines to "go green" more than ever before. One of the biggest engine improvements used to boost efficiency in recent years is the hybrid engine.

25. Hybrids were an obscure a decade ago, but now everyone knows how they work—an electric motor is partnered with a traditional gasoline engine in order to achieve high fuel economy numbers, but without the "range anxiety" of an electric engine, where the driver always wonders what will happen when a charge runs out.

(Source: http://auto.howstuffworks.com/10-improvements-in-engine-design11.html)

New Words and Expressions

cram	*v.*	填满，塞满
induction	*n.*	诱发
manifold	*n.*	（汽车引擎用于进气或排气）进气歧管
recoup	*v.*	补偿，偿还
shove	*v.*	推
supercharger	*n.*	增压器，增压机
turbocharger	*n.*	涡轮增压器

Comprehension Questions

1. The passage gives an example of turbocharged Mini Cooper S in order to explain ________.

   A. forced induction is more powerful without any increase in engine size

   B. forced induction consumes more fuel

   C. forced induction systems were used on aircraft engines long before

2. Which of the following is NOT an advantage of fuel injection?

   A. Increased fuel efficiency.

   B. Easier starting on cold days.

C. Higher industrial standard.

3. Fuel injection proves more favorable than carburettor mainly because ________.

   A. the fuel-injected engines are controlled by computer, making it easier to start

   B. engines of carburettors are more efficient and more responsive to changes in the throttle

   C. fuel injection systems are more costly to repair than carburettors

4. What does the underlined word "refinement" (line 1, para. 18) probably mean?

   A. 改革　　　　B. 修整　　　　C. 改善

5. Which of the following is NOT a reason for engines "going green" more than ever before?

   A. High gas prices.

   B. The falling fuel economy and emission standards.

   C. Rising awareness of environmental protection.

Grammar Coach

Imperative Sentence

祈使句（imperative sentence）是用于表达命令、请求、劝告、警告、禁止等的句子。祈使句在科技英语文本中的应用主要体现在产品使用说明、实验说明及技术的介绍与应用叙述等。下文将具体讲解祈使句结构及祈使句在科技英语中的应用。

1. 祈使句结构

（1）祈使句的主语通常被省略，多以动词原形开头，即"（please）动词原形+sth."。

例如：Turn on the light.

（2）在强调动词时，用助动词 do 引导祈使句，即"do+动词原形+sth."。

例如：Do be careful when crossing the street.

（3）祈使句的否定形式："(please)don't, never 等否定词+动词原形+sth."。

例如：Don't make a noise.

Never be late.

2. 祈使句在科技英语文本中的应用

（1）祈使句常用来表示条件："祈使句+and/or+完整句子"，and 或 or (else)后面的句子表示产生的结果。

例如：Press the button and the machine will operate by itself.

Keep off the high-tension transformer, or else you will get an electric shock.

（2）实验过程的描述及产品说明多用祈使句，不仅可以使语言表达简明扼要，同时也增强了语篇的示意和指向功能。

例如：Determine the volume of the specimen directly by sliding it into a graduated cylinder containing certain volume of water.

Make sure that no bubbles of air are trapped.

Note the total volume of water and specimen.

（3）科技英语中表示假设条件时，也常使用祈使句。常见的表明假设的动词有：assume、suppose、take、imagine 等。

例如：Imagine that there is an explosion in the reactor.

Assume there is no consumption of heat in this case.

Exercise

Translate the following sentences into English.

1. 请依照产品说明书使用，远离高温和火源，避免阳光直射。
2. 液体如不慎溢出，请立即擦拭干净，以免腐蚀仪表或真皮物件。
3. 请避免儿童接触，勿吞食；若不慎吞食，请立即就医。

Unit 12

Boiler

Text A

Boilers

1. Boilers are pressure vessels designed to heat water or produce steam, which can then be used to provide space heating or service water heating to a building. In most commercial building heating applications, the heating source in the boiler is a natural gas fired burner. Oil fired burners and electric resistance heaters can be used as well. Steam is preferred over hot water in some applications, including absorption cooling, kitchens, laundries, **sterilizers**, and steam driven equipment.

2. The key elements of a boiler include the burner, combustion chamber①, heat exchanger②, exhaust **stack**, and controls. Boiler accessories including the flue gas economizer are also commonly used as an effective method to recover heat from a boiler.

3. Boilers have several strengths that have made them a common feature of buildings. They have a long life, can achieve efficiencies up to 95% or greater, provide an effective method of heating a building, and in the case of steam systems, require little or no pumping energy. However, fuel costs can be considerable, regular maintenance is required, and if maintenance is delayed, repair can be costly. Boilers are often one of the largest energy users in a building. For every year a boiler system goes unattended, boiler costs can increase approximately 10%. Boiler operation and maintenance is therefore a good place to start when looking for ways to reduce energy use and save money.

---

① 燃烧室是一种用耐高温合金材料制作的燃烧设备，燃料或推进剂在其中燃烧生成高温燃气的装置。

② 换热器，也称热交换器或热交换设备，用来使热量从热流体传递到冷流体，以满足规定的工艺要求的装置，是对流传热及热传导的一种工业应用。

How Boilers Work

4. Both gas and oil fired boilers use controlled combustion of the fuel to heat water. The key boiler components involved in this process are the burner, combustion chamber, heat exchanger, and controls.

5. The burner mixes the fuel and oxygen together and, with the assistance of an **ignition** device, provides a platform for combustion. This combustion takes place in the combustion chamber, and the heat that it generates is transferred to the water through the heat exchanger. Controls regulate the ignition, burner firing rate, fuel supply, air supply, water temperature, steam pressure, and boiler pressure.

6. Hot water produced by a boiler is pumped through pipes and delivered to equipment throughout the building, which can include hot water **coils** in air handling units, service hot water heating equipment, and terminal units. Steam boilers produce steam that flows through pipes from areas of high pressure to areas of low pressure, unaided by an external energy source such as a pump. Steam utilized for heating can be directly utilized by steam using equipment or can provide heat through a heat exchanger that supplies hot water to the equipment.

Types of Boilers

7. Boilers are classified into different types based on their working pressure and temperature, fuel type, draft method, size and capacity, and whether they condense the water vapour in the combustion gases. Boilers are also sometimes described by their key components, such as heat exchanger materials or tube design. Two primary types of boilers include fire tube and water tube boilers. In a fire tube boiler, hot gases of combustion flow through a series of tubes surrounded by water. Alternatively, in a water tube boiler, water flows in the inside of the tubes and the hot gases from combustion flow around the outside of the tubes.

8. Fire tube boilers[①] are more commonly available for low pressure steam or hot water applications, and are available in sizes ranging from 500,000 to 75,000,000 Btu[②] input. Water tube boilers[③] are primarily used in higher pressure steam applications and are used extensively for comfort heating applications. They typically range in size from 500,000 Btu to more than 20,000,000 Btu input.

9. Cast iron (sectional) boilers[④] are another type of boiler commonly used in commercial space heating applications. These types of boilers don't use tubes. Instead, they're built up from

---

① 火管锅炉又称锅壳式锅炉，燃料燃烧后产生的烟气在火筒或烟管中流过，对火筒或烟管外水、汽或汽水混合物加热。

② 英热单位，1 Btu=1 054.35 J。

③ 水管锅炉，即在锅筒外部设水管受热面，高温烟气在管外流动放热，水在管内吸热。

④ 铸铁锅炉，由铸铁铸造成的锅片组装而成的热水锅炉。

cast iron sections that have water and combustion gas passages. The iron castings are bolted together, similar to an old steam radiator. The sections are sealed together by **gaskets**. They're available for producing steam or hot water, and are available in sizes ranging from 35,000 to 14,000,000 Btu input.

10. Cast iron boilers are advantageous because they can be assembled on site, allowing them to be transported through doors and smaller openings. Their main disadvantage is that because the sections are sealed together with gaskets, they are prone to leakage as the gaskets age and are attacked by boiler treatment chemicals.

Safety Issues

11. All combustion equipment must be operated properly to prevent dangerous conditions or disasters from occurring, causing personal injury and property loss. The basic cause of boiler explosions is ignition of a combustible gas that has accumulated within the boiler. This situation could arise in a number of ways, for example fuel, air, or ignition is interrupted for some reason, the flame **extinguishes**, and combustible gas accumulates and is reignited. Another example is when a number of unsuccessful attempts at ignition occur without the appropriate **purging** of accumulated combustible gas.

12. Boiler safety is a key objective of the National Board of Boiler and Pressure Vessel Inspectors. This organization reports and tracks boiler safety and the number of incidents related to boilers and pressure vessels each year. Their work has found that the number one incident category resulting in injury was poor maintenance and operator error. This stresses the importance of proper maintenance and operator training.

13. Boilers must be inspected regularly based on manufacturer's recommendations. Pressure vessel **integrity**, checking of safety relief valves①, water cutoff devices and proper float operation, **gauges** and water level indicators should all be inspected. The boiler's fuel and burner system requires proper inspection and maintenance to ensure efficient operation, heat transfer and correct flame detection.

（Source: https://betterbricks.com/sites/default/files/operations/om_of_boilers_final.pdf）

New Words and Expressions

coil *n.* 线圈

extinguish *v.* 熄灭

① 安全泄放阀是通过调节将进口压力减至某一需要的出口压力，并依靠介质本身的能量，使出口压力自动保持稳定的阀门。

gasket	*n.*	密封垫，密封圈
gauge	*n.*	测量，评估
integrity	*n.*	完整，完好
ignition	*n.*	燃烧，着火
purging	*n.*	净化，清除
sterilizer	*n.*	消毒器
stack	*n.*	烟囱

Comprehension Checks

Work in pairs and discuss the following questions. Then share your ideas with the rest of the class.

1. Could you state some of the key elements of boilers? (para.2)
2. In what way does the burner serve as a platform for combustion? (para.5)
3. Which type of boiler is suitable for low pressure steam or hot water applications? (para.8)
4. Could you briefly explain the advantages and disadvantages of cast iron boilers? (para.10)
5. What might be causing boiler explosion and how could that be prevented? (para.11)

Vocabulary Tasks

I. Fill in the blanks with the words given below. Change the forms where necessary.

boiler	maintenance	integrity	ignition	draft
transfer	extinguish	combustion	valve	flame

1. The gasoline vaporizes and mixes with the air to form a highly ________ mixture.
2. The heat from the ________ was so intense that road melted.
3. The company is responsible for ________ this equipment.
4. Some form of encryption is needed to protect confidentiality and ________ of data.
5. The people who stood on the avenue tried to ________ the fire, but it's of no avail.
6. The bombs ________ a fire which destroyed some 60 houses.
7. The pressure in the cylinders would go up in proportion to the ________ pressure.
8. Open doors can provide enough ________ so that a fire can run wild.
9. The presence or absence of clouds can have an important impact on heat ________.
10. A spurt of diesel came from one ________ and none from the other.

II. Match each word with the correct definition.

1. boiler	a. the state of being whole and not divided
2. draft	b. an instrument for measuring the amount or level of sth.
3. integrity	c. a flat piece of rubber, etc. placed between two metal surfaces in a pipe or an engine to prevent steam, gas or oil from escaping
4. extinguish	d. a container in which water is heated to provide hot water and heating in a building or to produce steam in an engine
5. ignition	e. a machine or piece of equipment that you use to make objects or substances completely clean and free from bacteria
6. sterilizer	f. the action of starting to burn or of making sth. burn
7. gauge	g. to make a fire stop burning or a light stop shining
8. gasket	h. a current of air (usually coming into a chimney or room or vehicle)

III. Below is an introduction to boiler. Put the sentences in the right order.

(   ) The heated or vaporized fluid exits the boiler for use in various processes or heating applications.

(   ) A boiler is a closed vessel in which water or other fluid is heated.

(   ) In North America, the term "furnace" is normally used if the purpose is not to boil the fluid.

(   ) The fluid does not necessarily boil.

(   ) For example, it can be used for water heating, central heating, boiler-based power generation, cooking, and sanitation.

Text B

Advances in Boiler Construction and Design

1. Contractors, architects, engineers, and others who specify commercial and industrial boilers continually seek systems that are more cost effective, energy efficient, and environmentally friendly. The boiler industry has responded by making significant advances in both construction and design, with new systems offering greater efficiencies, cutting-edge technologies, advanced controls, and the ability to integrate with renewable energy.

Condensing Boilers Gaining Momentum

2. The boiler market has been slowly shifting from standard, non-condensing boilers to more efficient condensing units over the last couple of decades. Reasons for the switch to condensing technology include a desire for operating efficiencies that result in lower operating

costs, increased awareness of energy efficient products, and the growing number of governmental policies and incentive programs.

3. Energy-efficient condensing boilers first appeared in the U.S. market in the late 1990s and their sales have grown **exponentially**. The movement accelerated when the energy crisis several years ago spiked fuel costs. The shift toward green building design and Leadership in Energy and Environmental Design (LEED)① **certifications** for installing high-efficiency equipment has also driven the trend.

4. Many predicted that condensing boilers would make up half of the market by 2016, but the trend has been slower than **anticipated**. This delay is attributed to a couple of factors. For one, North American economies are not growing as quickly as anticipated, making it more difficult for some businesses and homeowners to approve an investment in more advanced, pricier boiler technology. Additionally, federal rebates, which play a major role in encouraging installation of condensing technology, are no longer available. Some state and local **municipalities** offer rebates to help **offset** first boiler installation costs, but the lack of federal incentives has slowed growth.

5. Though the market today is still dominated by cast iron non-condensing boilers — which make up about 60 percent of installations — it's likely that condensing boilers will continue to experience growth and play an even larger role in the hydronic heating industry by 2017.

Greater Efficiency Key Driver

6. With an operating life measured in decades, purchasing a new or replacement condensing or non-condensing boiler is a significant decision. Most specifying engineers today seek the most energy efficient heating design.

7. Since September 2012, the minimum boiler efficiency requirement in the U.S. is 82 percent for gas hot water boilers and 84 percent for oil hot water boilers. Some **jurisdictions** in the U.S. have building or energy codes that require the reporting of the efficiency ratings of boilers and other heating products. Standard efficiency is anything below 90 percent annual fuel utilization efficiency (AFUE) while more than 90 percent is considered high efficiency. To reach levels above 90 percent, specifiers should consider installing a condensing boiler, as these units offer the greatest thermal efficiency. Typical aluminum or **stainless** steel condensing models offer efficiency greater than 90 percent AFUE.

8. A condensing boiler extracts additional heat from the exhaust gases by condensing its hot water vapour to liquid water, thus recovering its **latent** heat of vaporization. By capturing

---

① LEED是一个评价绿色建筑的工具，由美国绿色建筑协会建立，并于2003年开始推行，在美国的一部分州和其他一些国家已被列为法定强制标准。

some of the waste heat, the condensing boiler heat exchanger can be up to 10 percent more efficient than a conventional boiler operating in the proper conditions.

9. Some boiler manufacturers currently add a secondary heat exchanger or **recuperator** to improve the efficiency of conventional, non-condensing boilers. In this case, a secondary heat exchanger recovers the latent exhaust heat from the flue gases, condensing water vapour in the process. In theory, you can increase the efficiency of a standard cast iron boiler by up to 10 percent to achieve higher efficiency, lower energy costs, and qualify for utility rebates. Maintaining optimum boiler efficiency is critically important to minimizing carbon dioxide emissions, conserving fuel resources, and lowering building operating costs.

Improvements in Boiler Maintenance

10. The ease of cleaning and maintenance of condensing boilers varies based on the type of technology and material design. Stainless steel condensing boilers feature either a water tube or fire tube heat exchanger. The water tube stainless steel heat exchanger consists of water inside a spiral tube with the flue gases on the outside. These units can be more difficult to clean due to the spiral design.

11. The fire tube, on the other hand, features the flue gases on the inside and water on the outside. These are typically straight tubes and a facility manager can easily access them by opening the lid and cleaning them. As a result, fire tubes are becoming more popular in stainless steel condensing technology than water tubes.

12. Aluminum condensing heat exchangers feature separate sections bolted together and include a clean out plate. Inside the heat exchanger are pins to increase the heat transfer area. Maintenance personnel simply remove the front plate to access the inside of the heat exchanger. The areas between the pins can be then easily cleaned and flushed with water.

13. When HVAC equipment fails, it can be a building owner's and facility manager's worst nightmare. Heating units are typically down when they are running and needed the most. Also, there often isn't a repair budget in place because the failure wasn't anticipated. Therefore, facility managers today seek preventative maintenance options to avoid an expensive critical failure.

14. One new trend in boiler maintenance that meets this demand is predictive maintenance. With this new advancement, maintenance personnel can receive a predictive analysis of what may happen to a boiler over a specific length of time such as the next 6 to 12 months.

15. Manufacturers have developed this technology by factoring a certain set of **parameters** and based on changes in those factors can predict when specific maintenance procedures — such as cleaning tubes or a heat exchanger — should occur to prevent **catastrophic** failure.

(Source: http://www.achrnews.com/articles/131381-advances-in-boiler-construction-and-design)

New Words and Expressions

anticipate	*v.*	预见，预料
catastrophic	*adj.*	灾难的，惨重的
certification	*n.*	证书
exponentially	*adv.*	迅速增长地，成指数倍增长地
jurisdiction	*n.*	管辖权，司法权
latent	*adj.*	潜在的
municipality	*n.*	自治区，自治市
offset	*n.*	抵消，补偿
parameter	*n.*	参数
recuperator	*n.*	回流换热器，蓄热器
stainless	*adj.*	不锈的

Comprehension Questions

1. Which of the following does not contribute to the switching to condensing technology?
   A. Increased awareness of energy efficient products.
   B. The growing number of governmental policies and incentive programs.
   C. Higher operating costs.
2. What does the underlined word "extracts" (line 1, para. 8) probably mean?
   A. 释放 B. 吸收 C. 转化
3. Which of the following statement is TRUE about the boiler market?
   A. The market is dominated by condensing boilers.
   B. The condensing boilers will play a more important role in the future.
   C. Non-condensing boilers will continue growing.
4. Why do manufacturers add a secondary heat or recuperator?
   A. Because they want to lower the cost of using boilers.
   B. Because they want to improve the efficiency of condensing boilers.
   C. Because they want to increase the efficiency of conventional, non-condensing boilers.
5. What advantage do fire tubes have over water tubes?
   A. They can be cleaned easily by opening the lid.
   B. They are typically curl tubes.
   C. They have water on the inside and gases on the outside.

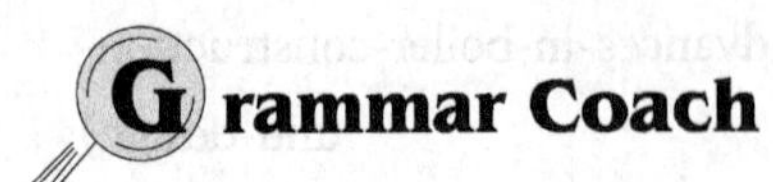

Grammar Coach

Compound Sentences

复合句（compound sentence）在科技英语阅读及写作中应用广泛。因为科技英语文本内容的理论性较强，经常使用复合句阐释一些比较复杂的概念，因此掌握复合句的特点及用法对于科技英语的学习十分重要。下文将主要对三类常见的主从复合句进行具体的介绍。

1. 定语从句

定语从句是由关系词引导的，在复合句中起形容词作用，修饰主句中的名词或代词的从句。被修饰的名词或代词称为先行词。定语从句一般跟在先行词之后，由关系代词及关系副词引导。举例如表 12–1 所示。

表 12–1　定语从句举例

引导词	句式特点	例句
关系代词	who、whom、that、which、whose、as 在从句中充当主语、宾语、表语、定语等	A body whose velocity is not constant is said to be accelerated. Einstein, who found the Theory of Relativity, won the Nobel Prize in 1921.
关系副词	where、when、why 在从句中充当状语	This is the reason why we should protect the earth. We should protect our planet where we live.

2. 名词性从句

名词性从句是在主句中起名词作用，是由连词引导的从句，主要包括主语从句、宾语从句、表语从句、同位语从句等。引导名词性从句的连词包括从属连词、连接代词及连接副词。举例如表 12–2 所示。

表 12–2　名词性从句举例

引导词	句式特点	例句
从属连词	that、whether、if 在从句中不充当任何成分	That price will go up is certain. Whether a transformer is set-up or step-down refers to voltage level.
连接代词	what、whatever、who、whoever、whom、whose、which 在从句中充当主语、宾语、表语、定语等	Who will be in charge of the department has not been decided yet. Whoever breaks the rule must be punished.
连接副词	when、where、how、why 在从句中充当状语	When the resource will be exploited remains unknown. This example points out why it's necessary to check solutions in the original equation.

3. 状语从句

状语从句又被称为副词性从句，在主句中作状语，用来修饰谓语、非谓语动词、定语、状语或整个句子。状语从句总共分为九类，每个状语从句都由相应的引导词或词组引导。举例如表 12–3 所示。

表 12–3　状语从句举例

从句分类	常用引导词或词组	例句
时间状语从句	when, while, as, whenever, till, until, not…until, before, after, since, once, immediately, directly, instantly, as soon as, the moment, the instant, the minute, every time, each time, no sooner…than, hardly…when	A body at rest will not move till the force impels it. As radio travels along the surface of the earth, they will lose energy.
地点状语从句	where, wherever	Air will be heavily polluted where there are factories. Red Cross is expected to send help wherever there is human suffering.
原因状语从句	because, as, since, for, now that, seeing that, considering that, in that	Now that everyone is here, let's begin our conference. The higher income tax is harmful because it discourages people from trying to earn more.
条件状语从句	if, unless, in case, as, so long as, provided that, providing, supposing, suppose that, on condition that, only if, if only	We will begin our project as long as the president agrees. Provided that there is no opposition, we will hold the meeting.
目的状语从句	so that, in order that, lest, for fear that, in case	The magnet is made in the shape of horseshoe so that it will be as strong as possible. The boss asked his secretary to hurry up with the letters so that he could sign them as soon as possible.
结果状语从句	so…that, such…that	The oil resource is so scarce that we should exploit it reasonably.
比较状语从句	as…as	Coal is not as clean as natural gas.
方式状语从句	as if, as though, as, just as	He behaved as if he were the boss.
让步状语从句	though, although, as, even if, even though, while, whereas, whatever, whoever, whichever, however, whenever, no matter what, no matter who, no matter which, no matter how, no matter when, whether…or not	All bodies emit heat rays no matter how low or high their temperature are. Although a nucleus is small, it consists of smaller units.

Exercise

Translate the following sentences into English.

1. 我们唯一需要做的事情就是调整阻力。（定语从句）
2. 许多国家都建立了国家公园，从而使动植物得到保护。（定语从句）
3. 目前无法预测何时才能取得这些重大的突破。（名词性从句）
4. 这个商店的与众不同之处就在于它提供了更多的个人服务。（名词性从句）
5. 球遇到障碍物才会停下来。（状语从句）

Unit 13

Turbine

Text A

Turbines

What is a Turbine?

1. A windmill is the simplest kind of turbine: a machine designed to capture some of the energy from a moving fluid (a liquid or a gas) so it can be put to use. As the wind blows past a windmill's sails, they rotate, removing some of the wind's kinetic energy[①] (energy of movement) and converting it into mechanical energy[②] that turns heavy, rotating stones inside the mill. The faster the wind blows, the more energy it contains; the faster the sails **spin,** the more energy is supplied to the mill. Adding more sails to the windmill or changing their design so they catch the wind better can also help to capture more of the wind's energy. Although you may not realize it, the wind blows just a bit more slowly after it's passed by a windmill than before — it's given up some of its energy to the mill!

2. The key parts of a turbine are a set of blades that catch the moving fluid, a **shaft** or **axle** that rotates as the blades move, and some sort of machine that's driven by the axle. In a modern wind turbine, there are typically three **propeller**-like blades attached to an axle that powers an electricity **generator**. In an ancient waterwheel, there are wooden **slats** that turn as the water flows under or over them, turning the axle to which the wheel is attached and usually powers some kind of milling machine.

3. Turbines work in two different ways described as **impulse** and reaction — terms that

---

① 动能是物体由于运动而具有的能量。

② 机械能是动能与部分势能的总和，这里的势能分为重力势能和弹性势能。

are often very confusingly described (and sometimes completely muddled up) when people try to explain them. So what's the difference?

Impulse Turbines

4. In an impulse turbine, a fast-moving fluid is fired through a narrow **nozzle** at the turbine blades to make them spin around. The blades of an impulse turbine are usually bucket-shaped so they catch the fluid and direct it off at an angle or sometimes even back the way it came (because that gives the most efficient transfer of energy from the fluid to the turbine). In an impulse turbine, the fluid is forced to hit the turbine at high speed.

5. Imagine trying to make a wheel like this turn around by kicking soccer balls into its paddles. You'd need the balls to hit hard and bounce back well to get the wheel spinning — and those constant energy impulses are the key to how it works. The law of conservation of energy① tells us that the energy the wheel gains, each time a ball **strikes** it, is equal to the energy that the ball loses — so the balls will be traveling more slowly when they bounce back. Also, Newton's second law of motion② tells us that the momentum gained by the wheel when a ball hits it is equal to the momentum lost by the ball itself; the longer a ball touches the wheel, and the harder (more forcefully) it hits, the more momentum it will transfer.

6. Water turbines are often based around an impulse turbine (though some do work using reaction turbines). They're simple in design, easy to build, and cheap to maintain, not least because they don't need to be contained inside a pipe or housing (unlike reaction turbines).

Reaction Turbines

7. In a reaction turbine, the blades sit in a much larger volume of fluid and turn around as the fluid flows past them. A reaction turbine doesn't change the direction of the fluid flow as **drastically** as an impulse turbine: it simply spins as the fluid pushes through and past its blades. Wind turbines are perhaps the most familiar examples of reaction turbines.

8. If an impulse turbine is a bit like kicking soccer balls, a reaction turbine is more like swimming — in reverse. Let me explain! Think of how you do freestyle (front crawl) by hauling your arms through the water, starting with each hand as far in front as you can reach and ending with a "follow through" that throws your arm well behind you. What you're trying to achieve is to keep your hand and forearm pushing against the water for as long as possible, so you transfer as much energy as you can in each stroke. A reaction turbine is using the same

① 能量守恒定律，即热力学第一定律，是指在一个封闭系统的总能量保持不变，其中总能量一般来说已不再只是动能与势能之和。

② 牛顿第二运动定律，简称牛顿第二定律，物体随时间变化之动量变化率和所受外力之和成正比，即动量对时间的一阶导数等于外力之和。

idea in **reverse**: imagine fast-flowing water moving past you so it makes your arms and legs move and supplies energy to your body! With a reaction turbine, you want the water to touch the blades smoothly, for as long as it can, so it gives up as much energy as possible. The water isn't hitting the blades and bouncing off, as it does in an impulse turbine: instead, the blades are moving more smoothly, "going with the flow".

New Words and Expressions

axle	*n.*	车轴，车桥
drastically	*adv.*	彻底地
generator	*v.*	发电机
impulse	*n.*	冲量，冲击，推动力
nozzle	*n.*	喷嘴
propeller	*n.*	螺旋桨，推进器
reverse	*n.*	反转，颠倒
shaft	*n.*	柄，轴
slat	*n.*	板条，狭板
strike	*v.*	打击，敲击

Comprehension Checks

Work in pairs and discuss the following questions. Then share your ideas with the rest of the class.

1. What are the main components of a turbine? (para. 2)
2. Why are the blades of an impulse turbine often bucket-shaped? (para.4)
3. How might the law of conservation of energy explain the working principle of the impulse turbine? (para. 5)
4. What are some of the differences between impulse turbines and reaction turbines? (para. 7 & para. 8)
5. In what way could reaction turbines be compared to swimming? (para. 8)

Vocabulary Tasks

I. Fill in the blanks with the words given below. Change the forms where necessary.

subtly	extract	divert	reverse	drastically
strike	conservation	shaft	impulse	generator

1. The local economy is overwhelmingly dependent on oil and gas ________.
2. The government is trying to ________ more public funds from west to east.
3. The domestic political situation changed ________.
4. The company, New England Electric, burns coal to ________ power.
5. Hamilton Oil announced that it had ________ oil in the Liverpool Bay area of the Irish Sea.
6. Herbs have been used for centuries to endow a whole range of foods with ________ flavors.
7. The ________ of falling water turns the waterwheel.
8. The downward trend went into ________ and the scores started to creep up again.
9. Cheaper energy ________ techniques have been put into operation in the developed world.
10. In a machine, a ________ is a rod that turns round continually in order to transfer movement in the machine.

II. Match each word with the correct definition.

1. conservation	a. a machine for producing electricity
2. impulse	b. a metal bar that joins parts of a machine or an engine together, enabling power and movement to be passed from one part to another
3. reverse	c. to change or make something change from one form, purpose, system, etc. to another
4. generator	d. not immediately obvious or noticeable
5. subtly	e. the act of preventing something from being lost, wasted, damaged or destroyed
6. convert	f. a force or movement of energy that causes something else to react
7. shaft	g. a machine or an engine that receives its power from a wheel that is turned by the pressure of water, air or gas
8. turbine	h. to turn something the opposite way around or change the order of something around

III. Below is an introduction to turbine. Put the sentences in the right order.

( ) The work produced by a turbine can be used for generating electrical power when combined with a generator.

( ) A turbine is a rotary mechanical device that extracts energy from a fluid flow and converts it into useful work.

( ) Moving fluid acts on the blades so that they move and impart rotational energy to the rotor.

( ) Turbine is a turbo machine with at least one moving part called a rotor assembly, which is a shaft or drum with blades attached.

(　　) Early turbine examples are windmills and waterwheels.

Text B

Next-generation Wind Technology

1. The Wind Program works with industry partners to increase the performance and reliability of next-generation wind technologies while lowering the cost of wind energy. The program's research efforts have helped to increase the average capacity factor (a measure of power plant productivity) from 22% for wind turbines installed before 1998 to an average of 33% today, up from 30% in 2000. Wind energy costs have been reduced from over 55 cents (current dollars) per kilowatt-hour (kW • h) in 1980 to an average of 2.35 cents in the United States today.

2. To ensure future industry growth, the technology must continue to evolve, building on earlier successes to further improve reliability, increase capacity factors, and reduce costs. This essay describes the goal of the program's large wind technology research efforts and **highlights** some of its recent projects.

Prototype Development

3. Modern wind turbines are increasingly cost-effective and more reliable, and have scaled up in size to multi-megawatt power ratings. Since 1999, the average turbine generating capacity has increased, with turbines installed in 2014 averaging 1.9 MW of capacity. Wind Program research has helped **facilitate** this transition, through the development of longer, lighter rotor blades, taller towers, more reliable **drivetrains**, and performance-optimizing control systems. Furthermore, improved turbine performance has led to a more robust domestic wind industry that saw wind turbine technology exports grow from $16 million in 2007 to $488 million in 2014.

4. During the past two decades, the program has worked with industry to develop a number of **prototype** technologies, many of which have become commercially viable products. One example is the GE Wind Energy 1.5-MW (megawatt) wind turbine. Since the early 1990s, the program worked with GE and its **predecessors** to test components such as blades, generators, and control systems on generations of turbine designs that led to GE's 1.5-MW model. The GE 1.5 constitutes approximately half of the nation's installed commercial wind energy fleet, and is a major competitor in global markets.

Component Development

5. The program works with industry partners to improve the performance and reliability of system components. Knight and Carver's Wind Blade Division in National City, California, worked with researchers at the Department of Energy's Sandia National Laboratories to develop an **innovative** wind turbine blade that has led to an increase in energy capture by 12%. The most distinctive characteristic of the Sweep Twist Adaptive Rotor (STAR) blade is a gently curved tip, which, unlike the vast majority of blades in use, is specially designed to take maximum advantage of all wind speeds, including slower speeds.

6. To support the development of more reliable gearboxes, the program has worked with several companies to design and test innovative drivetrain concepts. Through the support of $47 million in DOE① funding, the nation's largest and one of the world's most advanced wind energy testing facilities was opened at Clemson University② to help speed the deployment of next generation energy technology, reduce costs for manufacturers and boost global competitiveness for American companies. The Clemson facility is equipped with two testing bays—for up to 7.5-MW and 15- MW drivetrains, respectively. The Clemson facility also features a **grid** simulator that **mimics** real-world conditions, helping researchers better study the interactions between wind energy technologies and the U.S. power grid.

Utility-scale Research Turbine

7. In 2009, the program installed a GE 1.5-MW wind turbine at the National Wind Technology Center (NWTC) located on the National Renewable Energy Laboratory campus in Boulder, Colorado. This turbine was the first large-scale wind turbine fully owned by DOE and serves as a platform for research projects aimed at improving the performance of wind technology and lowering the costs of wind energy. The NWTC is now **collaborating** with Siemens Energy to conduct aerodynamic field experiments on a 2.3-MW wind turbine. These experiments <u>utilize</u> **sonic** as well as conventional anemometers and wind vanes on the NWTC's 135 meter **meteorological** tower to measure characteristics such as inflow, turbine response, and wind wake. The data gained from these experiments will provide new insights into multi-megawatt turbine aerodynamic response, structural loading, power production, and fatigue life that can be used to increase reliability and performance. The research being done at the NWTC complements DOE's Atmosphere to Electrons (A2e) initiative that targets significant reductions in the cost of wind energy through an improved understanding of the complex physics

---

① 美国能源部是美国联邦政府的一个下属部门，主要负责美国联邦政府能源政策制定，能源行业管理，能源相关技术研发、武器研制等。

② 克莱姆森大学，是美国二十所顶尖的公立大学之一，位于美国南卡罗来纳州克莱姆森市。

governing wind flow into and through wind farms.

International Collaboration

8. As a member of the International Energy Agency (IEA)[①] Wind Energy Executive Committee, the program supports international wind energy research efforts by participating in 12 areas of wind energy research. The program's participation in these international research efforts provides U.S. researchers an opportunity to collaborate with international experts in wind energy, exchange recent technical and market information, and gain valuable feedback for the U.S. Industry.

New Words and Expressions

collaborate	*v.*	合作，协作
drivetrain	*n.*	动力传动系统
facilitate	*v.*	帮助，促进
grid	*n.*	网格，输电网
highlight	*v.*	强调，突出
innovative	*adj.*	创新的
meteorological	*adj.*	气象的，气象学的
mimic	*v.*	模拟，模仿
predecessor	*n.*	前身
sonic	*adj.*	声波的，能发声音的

Comprehension Questions

1. Which of the following does not contribute to industry growth in the future ?
   A. Reducing the cost of technology.
   B. Maintaining the average capacity factor.
   C. Further improving reliability.
2. Why does the author give the example of GE in the 4th paragraph?
   A. To explain why it is a major competitor in global market.
   B. To justify that GE 1.5 consists nearly half of the nation's installed commercial wind energy fleet.

① 国际能源机构，经济合作与发展组织的辅助机构之一，1974年11月成立，现有成员国21个，总部设在法国巴黎。它的宗旨是： 协调各成员国的能源政策，减少对进口石油的依赖，在石油供应短缺时建立分摊石油消费制度，促进石油生产国与石油消费国之间的对话与合作。

C. To prove that many prototype technologies have become commercially viable products.

3. In which aspect is the STAR blade different from vast majority of blades in use?

A. It can make the most of all wind speed.

B. It is designed with curved edge.

C. It can capture more energy.

4. What does the underlined word "utilize" (line 6, para. 7) probably mean?

A. 优化　　B. 利用　　C. 简化

5. How would the participation of the Wind Energy Executive Committee benifit the US in their research efforts?

A. Optimizing their wind energy technology structure.

B. Being more competitive in the international energy market.

C. Enhancing the cooperation with international experts in wind energy.

Grammar Coach

Subject-Verb Agreement

主谓一致（subject-verb agreement）指的是谓语动词在人称和数上要和主语保持一致。主谓一致主要包括语法一致、意义一致和就近一致。下文将从这三个方面对主谓一致进行讲解。

1. 语法一致

语法一致即谓语动词在单复数形式上要和主语保持一致。主语为单数形式，谓语动词也为单数形式；主语为复数形式，谓语动词也为复数形式。

（1）当可数名词单数、不可数名词、代词（第三人称单数）、动词不定式、动名词或从句作主语时，谓语动词用单数形式；当可数名词复数作主语时，谓语动词用复数形式。

例如：Natural resource is very scarce and limited.

Explorations of oil have been made in some regions of the country.

（2）主语为可数名词单数或代词（第三人称单数）时，若后面有 with、together with、as well as、like、but、except 等单词或短语，谓语动词用单数形式；若主语为复数，谓语动词用复数形式。

例如：Oil, together with natural gas, is of vital importance to the development of China.

Nobody but Jim and Terry was on the playground.

（3）当被 every、each、no、many a、more than one 等限定词修饰的名词或由 some、any、no、every 构成的复合不定代词作主语时，谓语动词用单数形式。

例如：Everything around us is matter.

Each of us is responsible for protecting the environment.

（4）当“plenty of/a lot of/lots of/分数、百分数+名词”作主语时，谓语动词的数应与该结构中的名词保持一致。

例如：Plenty of water has been used to irrigate.

30% of the population live in the countryside.

（5）在定语从句中，若关系代词 that、who、which 作主语，谓语动词的数应与句中先行词的数一致。

例如：Coal Gangue is a waste from the raw coal which pollutes the environment seriously.

Radon is a gas that comes from the radioactive decay of radium in rocks.

（6）在倒装句中，谓语动词的数应与主语保持一致。

例如：Here comes to the point.

Such are the facts.

2. 意义一致

意义一致就是谓语动词要和主语意义上的单复数保持一致。

（1）由 and 连接合成主语，若两个主语表示两个概念，谓语动词用复数形式；若两个主语指同一人的同一事物，谓语动词用单数形式。但是由 both…and 连接的合成成分作主语，谓语动词用复数形式。

例如：The writer and composer is going to deliver a speech.

Both coal and oil are primary energy.

（2）在集体名词作主语时，若是整个集体的概念，谓语动词用单数形式；若强调的是集体中的个体，谓语动词用复数形式。集体名词有：class、family、audience、population 等。但是当 cattle、people 和 police 作主语时，谓语动词用复数形式。

例如：The population is increasing at about 6% per year.

All the family enjoy skiing.

（3）当表示时间、距离、重量、长度和价值的名词作主语时，尽管该类词是复数形式，也作为一个整体来看待，谓语动词用单数形式。

例如：Three years has passed.

One hundred miles is a long distance.

（4）-s 结尾的学科名词，如 maths、physics、politics 等，以及 news、the United States 等名词或短语作主语时，尽管形式上是复数，但意义上是单数，谓语动词用单数形式。

例如：The United States imports sugar from Cuba.

News of a serious road accident is just coming in.

3. 就近一致

就近一致就是谓语动词要和靠近它的主语部分保持一致。

（1）由 or、nor、either...or、neither...nor、not only... but (also)等连接的并列主语。

例如：Either you or I am going to be in charge of this matter.

Not only she but also he is to be blamed for the matter.

（2）there be 句型。

例如：There is a dictionary and some books for you.

There are some books and a dictionary for you.

Exercise

Translate the following sentences into English.

1. 中国约 90%的一次能源都是由本国提供的。
2. 全镇的居民都出席了集会。
3. 无论他们还是我们都不是完全正确的。（neither…nor…）
4. 天然气发电比煤便宜。
5. 目前在巴基斯坦存在能源危机。（there be）
6. 每个州对回收再利用的物品和时间都制定了不同的规章和条例。

Unit 14

Air Conditioning and Refrigeration System

Text A

How Air Conditioners Work

1. The first modern air conditioning system was developed in 1902 by a young electrical engineer named Willis Haviland Carrier①. It was designed to solve a humidity problem that at the Sackett-Wilhelms Lithographing and Publishing Company in Brooklyn, paper stock at the plant would sometimes absorb moisture from the warm summer air, making it difficult to apply the layered inking techniques of the time. Carrier treated the air inside the building by blowing it across chilled pipes. The air cooled as it passed across the cold pipes, and since cool air can't carry as much moisture as warm air, the process reduced the humidity in the plant and **stabilized** the moisture content of the paper. Reducing the humidity also had the side benefit of lowering the air temperature — and a new technology was born.

2. Carrier realized he'd developed something with far-reaching potential, and it wasn't long before air conditioning systems started popping up in theaters and stores, making the long, hot summer months much more comfortable.

3. The actual process air conditioners use to reduce the ambient air temperature in a room is based on a very simple scientific principle. The rest is achieved with the application of a few clever mechanical techniques. Actually, an air conditioner is very similar to another appliance in your home — the refrigerator. Air conditioners don't have the exterior housing a refrigerator

① 威利斯·哈维兰·卡里尔：美国工程师及发明家，是现代空调系统的发明者、开利空调公司的创始人，因其对空调行业的巨大贡献，被后人誉为“空调之父”。

relies on to insulate its cold box. Instead, the walls in your home keep cold air in and hot air out.

Air-conditioning Basics

4. Air conditioners use refrigeration to chill indoor air, taking advantage of a remarkable physical law: when a liquid converts to a gas (in a process called phase conversion), it absorbs heat. Air conditioners exploit this feature of phase conversion by forcing special chemical **compounds** to evaporate and **condense** over and over again in a closed system of coils.

5. The compounds involved are **refrigerants** that have properties enabling them to change at relatively low temperatures. Air conditioners also contain fans that move warm interior air over these cold, refrigerant-filled coils. In fact, central air conditioners have a whole system of ducts designed to **funnel** air to and from these **serpentine**, air-chilling coils.

6. When hot air flows over the cold, low-pressure evaporator coils, the refrigerant inside absorbs heat as it changes from a liquid to a gaseous state. To keep cooling efficiently, the air conditioner has to convert the refrigerant gas back to a liquid again. To do that, a compressor puts the gas under high pressure, a process that creates unwanted heat. All the extra heat created by compressing the gas is then evacuated to the outdoors with the help of a second set of coils called condenser coils, and a second fan. As the gas cools, it changes back to a liquid, and the process starts all over again. Think of it as an endless, elegant cycle: liquid refrigerant, phase conversion to a gas/heat absorption, compression and phase transition back to a liquid again.

7. It's easy to see that there are two distinct things going on in an air conditioner. Refrigerant is chilling the indoor air, and the resulting gas is being continually compressed and cooled for conversion back to a liquid again.

The Parts of an Air Conditioner

8. The biggest job an air conditioner has to do is to cool the indoor air. That's not all it does, though. Air conditioners monitor and regulate the air temperature via a thermostat. They also have an onboard filter that removes airborne particulates from the circulating air. Air conditioners function as **dehumidifiers**. Because temperature is a key component of relative humidity, reducing the temperature of a volume of humid air causes it to release a portion of its moisture. That's why there are drains and moisture-collecting pans near or attached to air conditioners, and why air conditioners discharge water when they operate on humid days.

9. Still, the major parts of an air conditioner manage refrigerant and move air in two directions: indoors and outside.

10. The cold side of an air conditioner contains the evaporator and a fan that blows air over the chilled coils and into the room. The hot side contains the compressor, condenser

and another fan to vent hot air coming off the compressed refrigerant to the outdoors. In between the two sets of coils, there's an expansion valve. It regulates the amount of compressed liquid refrigerant moving into the evaporator. Once in the evaporator, the refrigerant experiences a pressure drop, expands and changes back into a gas. The compressor is actually a large electric pump that pressurizes the refrigerant gas as part of the process of turning it back into a liquid. There are some additional sensors, timers and valves, but the evaporator, compressor, condenser and expansion valve are the main components of an air conditioner.

11. Although this is a conventional setup for an air conditioner, there are a couple of variations you should know about. Window air conditioners have all these components mounted into a relatively small metal box that installs into a window opening. The hot air vents from the back of the unit, while the condenser coils and a fan cool and re-circulate indoor air. Bigger air conditioners work a little differently: Central air conditioners share a control **thermostat** with a home's heating system, and the compressor and condenser, the hot side of the unit, isn't even in the house. It's in a separate all-weather housing outdoors. In very large buildings, like hotels and hospitals, the exterior condensing unit is often mounted somewhere on the roof.

Energy Efficient Cooling Systems

12. Because of the rising costs of electricity and a growing trend to "go green", more people are turning to alternative cooling methods to spare their pocketbooks and the environment. Big businesses are even jumping on board in an effort to improve their public image and lower their overhead.

13. Ice cooling systems are one way that businesses are combating high electricity costs during the summer. Ice cooling is as simple as it sounds. Large tanks of water freeze into ice at night, when energy demands are lower. The next day, a system much like a conventional air conditioner pumps the cool air from the ice into the building. Ice cooling saves money, cuts pollution, eases the strain on the power grid and can be used alongside traditional systems. The downside of ice cooling is that the systems are expensive to install and require a lot of space. Even with the high startup costs, more than 3,000 systems are in use worldwide.

(Source: http://home.howstuffworks.com/ac.htm)

New Words and Expressions

compound	*n.*	混合物
condense	*v.*	凝结
dehumidifier	*n.*	除湿器

refrigerant *n.* 制冷剂
serpentine *adj.* 弯弯曲曲的
stabilize *v.* 稳定
thermostat *n.* 恒温器，自动调温器

Comprehension Checks

Work in pairs and discuss the following questions. Then share your ideas with the rest of the class.

1. How would cooled air solve the humidity problem of inking techniques? (para.1)
2. What might be the major difference between air conditioner and refrigerator? (para.3)
3. Could you explain the working principle of an air conditioner? (para.4)
4. Does refrigerant ever change its state? How and for what purpose? (para.6)
5. Why do air conditioners discharge water when it's operating on humid days? (para.8)
6. What advantages and downsides could ice cooling system present? (para.12-13)

Vocabulary Tasks

I. Complete the sentences below. The first letter of the missing word has been given.

1. C________ means that a continuous series of circular rings into which something such as wire or rope has been wound or twisted.
2. The cells have different sides, and one side must face the wall of the d________ and the other must face the inner channel.
3. R________ is chilling the indoor air, and the resulting gas is being continually compressed and cooled for conversion back to a liquid again.
4. Since the business computing environment changes from time to time, the separation of logical and physical data modeling will help s________ the logical models from phase to phase.
5. The evaporator and the c________ are both made out of copper refrigerant coils, and there is also some copper in the compressor.

II. Choose the correct answer from the three choices marked A, B, and C.

1. When the atria (心房) contract, the valves sandwiched between the atria and the ventricles (心室) open, and the blood in each atrium flows through its respective ________ down into a ventricle.

   A. valve　　B. bottle　　C. pile

2. Smart-phones are packed with sensors, measuring everything from the user's

location to the

________light.

A. bright B. ambient C. ambitious

3. The hot air ________ from the back of the unit, while the condenser coils and a fan cool and re-circulate indoor air.

A. leaks B. vents C. comes

4. Government lenders are offering "green mortgages" with lower interest rates to home buyers who ________ windows or install solar panels.

A. instruct B. make C. insulate

5. Water is a ________ containing the elements hydrogen and oxygen.

A. complex B. compound C. comprehensive

6. The steam will ________ on the inside of the microwave loosening any stuck on food.

A. condense B. conductive C. condemn

7. The Earth's atmosphere seeks a steady oxygen level much as a ________ hones in on a steady temperature.

A. thermostat B. bottle C. hot

8. Central air conditioners have a whole system of ducts designed to funnel air to and from these ________, air-chilling coils.

A. cold B. specific C. serpentine

Text B

Magnetic Air Conditioners: A High Tech Way of Keeping Cool

1. As the Earth's average temperature continue to rise, those of us in already-warm climates may find ourselves increasingly tempted to <u>crank up</u> the A/C[①]. But traditional air conditioners are huge consumers of energy. According to the U.S. Department of Energy, residential air conditioning accounts for 16 percent of all household electricity use in the United States.

2. One day in the not-too-distant future, magnetic air conditioners may help us keep our homes cooler inside without making temperatures hotter outside. A traditional air conditioner works by changing a liquid refrigerant to a gas (absorbing heat from the outside air in the process), then compressing and cooling the gas to convert it back to a liquid. Magnetic air conditioners take an entirely different approach to cooling, using magnets instead of compressors and refrigerants to cool the surrounding air.

---

① air conditioning，空调。

The Magnetocaloric Effect

3. Magnetic air conditioners are based on a phenomenon known as the **magnetocaloric** effect. Magnetic materials heat up when they are exposed to a magnetic field, then cool down when the field is removed. The magnetocaloric effect was first observed in iron samples by a German physicist named Emil Warburg[①] in 1881, but the change in temperature was too small for any practical application. In recent years, however, separate groups of researchers have developed magnetocaloric metal **alloys** that produce a significantly larger magnetocaloric effect at room temperature.

How will Magnetic Air Conditioners Work?

4. Magnetic air conditioners cool the air by quickly and repeatedly exposing a magnetocaloric material to a magnetic field. In one prototype designed by Astronautics Corporation of America[②] in conjunction with the U.S. DOE's Ames (Iowa) Laboratory[③], a wheel containing the rare-Earth element **gadolinium** spins through the field of a stationary magnet. As the disk spins, the gadolinium alloy heats up, then cools as it passes through a gap in the field, cooling the water that surrounds it as it makes its trip.

Environmentally Friendly Technology

5. In a magnetic air conditioner, the alloy acts as the "refrigerant", and plain old water provides the heat transfer, eliminating the need for the decidedly unfriendly **hydrochlorofluorocarbons** (HCFCs) used in traditional air conditioners. While the earliest magnetocaloric alloys were either toxic or prohibitively expensive, the latest magnetocaloric materials are cost-effective and environmentally safe.

6. Magnetic air conditioners do require electricity, but the motor that spins the disk containing the magnetocaloric alloy is expected to be much more efficient than the compressor required to run a traditional air conditioner. According to a 2011 article in *Scientific American*, Astronautics hopes to have a prototype by 2013 that uses just two-thirds as much electricity as a conventional air conditioner to provide the same amount of cooling.

Is There a Magnetic Air Conditioner in Your Future?

7. In addition to the researchers mentioned above (Astronautics and DOE), several private companies, universities and government agencies around the world are exploring

---

① 埃米尔·沃伯格，物理学家。

② 美国宇航公司。

③ 美国能源部艾姆斯国家实验室。

magnetic cooling technology for industrial and domestic applications including air conditioning, refrigeration and climate control.

8. The National Laboratory for Sustainable Energy at the Technical University of Denmark① runs its own "MagCool" project, while research at Penn State② and other U.S. universities has advanced the understanding of magnetocaloric principles, helping us understand why one material might cool more efficiently than another. In 2009, BASF③ and Delta Electronics④ announced a corporate partnership to develop new magnetocaloric cooling systems and "explore the opportunities of magnetocaloric power generation". But as much as we'd love to get our hands on this new technology today, commercial availability of magnetic air conditioners is still at least a few years away, and the first uses will most likely be industrial rather than residential.

(Source: http://home.howstuffworks.com/magnetic-air-conditioner.htm)

New Words and Expressions

alloy	*n.*	合金
gadolinium	*n.*	钆（金属名，缩写为 Gd）
hydrochlorofluorocarbon	*n.*	含氢氯氟烃（缩写为 HCFS）
magnetocaloric	*adj.*	磁致热的

Comprehension Questions

1. What does the underlined phrase "crank up" (line2, para.1) probably mean?
   A. 启动　　B. 生产　　C. 举起
2. What's the difference between magnetic air conditioners and traditional ones?
   A. The magnetic A/C boasts higher energy conversion efficiency.
   B. The cooling methods of the two are different.
   C. The traditional ones work by absorbing heat from the outside air.
3. Where was the magnetocaloric effect firstly discovered?
   A. In iron sample.　　B. In alloy sample.　　C. In aluminum sample.
4. Which of the following is TRUE about the working process of magnetic A/C?
   A. The magnetic A/C cools the air by quickly and repeatedly using compressors and refrigerants.
   B. A wheel containing the rare-Earth element gadolinium is used for cooling the air.

---

① 丹麦技术大学，又译丹麦科技大学，是世界顶尖的理工大学之一。
② 宾夕法尼亚州立大学是一所享誉世界的顶尖研究型大学，该校的学术科研能力一直位居世界前列。
③ 巴斯夫股份公司是德国的一家化工企业，也是世界最大的化工厂。
④ 台达集团创立于 1971 年，为全球电源管理与散热解决方案的领导厂商。

C. The magnetic A/C utilizes a new technology less pollutant than the traditional approach.

5. According to the passage, which of the following is NOT true?

A. The current magnetocaloric alloys have been improved to be inexpensive and environmentally safe.

B. There is no need for electricity in the operation of magnetic A/C.

C. The magnetic A/C cannot be currently applied to residential use because of its high cost.

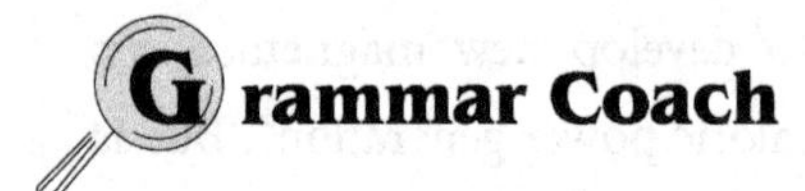

Grammar Coach

Parallel Structure

英语中的平行结构（parallel structure）是由两个或两个以上意义关联、语气一致、语法功能相同的同一类词、短语、分句等排列在一起构成的，运用得当可以使句子或语篇结构条理清晰，层次井然，节奏和谐，具有良好的文体和修辞效果。

科学技术领域中有许多平行不悖、前后衔接、相互照应的现象、过程、方法和规则，为了语言表述的严密性，科研人员在写作时通常会将若干并列成分纳入一体，归为一句，较多地运用平行结构，造成一种“突出”（foregrounding）。

1. 平行成分的组合

（1）连写。采用并列连词 and、but、as well as、or、or else、both…and、neither…nor、either…or、not only…but (also)、rather than 等，以及从属连词 than 连接两个对等的词或结构。

例如：Hydrostatics is fundamental to hydraulics, the engineering of equipment for storing, transporting and using fluids.

（2）分写。将平行项分行列出，予以突出。平行项开头首字母采用小写，项后标点符号为分号，句末为句点。

例如：The process includes:

a. leaching arsenic from a lead smelter copper dross dust;
b. fixing arsenic as an insoluble compound;
c. recovering other metal values.

（3）列表。当表示两个或两个以上相关项的并列关系，特别是有数据出现时，常用列表的形式，举例如表 14–1 所示。

表 14–1 列表举例

material	energy density/ ($MJ \cdot kg^{-1}$)	100 W light bulb time (1 kg)
wood	10.0	1.2 days
ethanol	26.8	3.1 days
coal	32.5	3.8 days
crude oil	41.9	4.8 days

2. 平行结构的分类

英语中的平行结构存在于各语言平面，适用于表达不同层次科技概念的需要。平行结构中的平行项之间不但在语义上有对应性，在语法结构和语法功能上有一致性，而且在词语的排列形式上保持匀称，如词、短语、分句等。

（1）词的平行。

例如： Small industrial enterprises producing iron and steel, chemical fertilizer, coal, machinery, cement, chemical fibers and paper and small hydroelectric power stations are all over the country.

（2）短语的平行。

例如： Power is the same in the following two cases:

Work done at the rate of 8ft.-1b. per second.

Work done at the rate of 200ft.-1b. per second.

（3）分句的平行。

例如： If indoors, describe physical treatment of walls, ceiling and floors, include a sketch showing location of source and room contents; if outdoors, include a sketch showing location of source with respect to surrounding terrain, including physical description of test environment.

3. 平行结构的词序安排

（1）符合逻辑顺序，即按事物本来次序或结构顺序加以安排，如工艺流程的先后、物质结构的由表及里等。

① 渐进（climax）：由浅入深、从低至高、由小至大、从轻到重，语义层层递进。

② 突降（anti-climax）：由多而少、由强而弱、由远及近、由深至浅。

（2）如果在内容上平行项的词序无关紧要，则要考虑语言形式，把较长、较复杂的平行项作为末项。

（3）根据句末中心的原则，把最重要的信息置于句末。

（4）按字母排列：国家名、地名、人名。

Exercise

Translate the following sentences and pay special attention to the parallel structures.

1. The manufacturers disclaim all responsibility for belongings or personal casualty caused by improper installation or misuse.
2. The vehicle will attain a top speed of 390 miles per hour at an altitude of 6,000 feet and a cruising speed of 350 miles per hour at 25,000 feet.
3. The LAMV is a vertical take-off and landing aircraft that can fly in a quick, quiet, and agile manner.
4. Multiple systems check fuel for quality and quantity and provide appropriate warning.
5. Their combined expertise in switching, radio and networking makes Ericsson a world leader in telecommunications.

Unit 15

Fuels and Combustion

Text A

Fuels, Combustion & Environmental Considerations in Industrial Gas Turbines

1. For economic and environmental reasons, it is important that gas turbines used in both industrial power generation and also oil & gas applications can burn a wide variety of fuels, with minimum impact on the environment.

Gas Turbine Fuel and Emission

2. Ensuring that fuel is provided and maintained at a high quality is key to delivering good operation in a modern gas turbine over long periods of time. However, it is not just fuel that is critical, it is also ensuring all fluids entering the GT are equally kept at a high standard, thus minimizing or eliminating all sources of **contaminants**.

3. Modern gas turbines operate at high temperatures, and use component designs and materials at the forefront of technology, but these are more susceptible to damage if contaminated fuel and air enter the GT through poor operating procedures. This article will consider the need to ensure good quality fluids enter the GT and review combustion technology which has moved forwards in achieving low emissions without resorting to wet **abatement** methods. Both conventional and low emissions technologies are discussed along with basic operating parameters associated with the fuels in question.

4. The fuel range used in GT applications is very wide with the choice based typically on availability and cost. In some cases, fuels may have little or no treatment, in others they may have "added value" which results in the high quality pipeline natural gas that provides the fuel

of choice for gas turbine OEMs and operators alike. Gas turbines can and do operate on a wide range of fuels, but the impact that such fuels may have on turbine life has to be recognized.

5. It is not a case of saying these fuels are acceptable, but understanding the details, such as the composition (hydrocarbon species in the case of a gaseous fuel, **inert** species, contaminants, water vapour, ...). Detailed analysis of the fuels is necessary to determine key parameters of the fluid, such as delivery, storage and conditioning as well as key features of the fuel itself, including lower heating value (LHV)①, Wobbe index②, **dew** point and density. Understanding all of these provides the OEM and users alike with indicators that the fuel entering the GT is suitable and can result in good operation across a wide range of loads and ambient conditions. It is also important to determine and understand the products of combustion and impact on the environment. Exhaust emissions are highly regulated in many parts of the world and even those areas that up until recently had no requirements have started to introduce standards or guidelines which need to be noted during the application assessment stage.

Types of Combustion Emissions Regulated

6. The U.S. Clean Air Act and the European Union (EU) Combustion Plant Directive are examples where limits on the worst polluting species from gas turbine plants are regulated. Consequently, low emissions became the **norm** and not the exception, especially for pollutants such as **nitrogen** oxide (NO_x), carbon **monoxide** (CO) and **unburnt** hydrocarbons (UHC).

7. In addition, the major gas turbine OEMs, along with a large number of oil & gas companies, have their own policies with regard to environmental **stewardship** and offer or specify low emission equipment even in locations where no formal legislation exists, or is set at a higher level.

8. The result of all of these drivers is to make the Dry Low Emissions or Dry Low NO_x (DLE/DLN) combustion system the primary combustion system of choice. In some cases, DLE/DLN is the only combustion system offered, especially on newer models.

Conventional Combustion

9. Conventional combustion, also referred to as diffusion flame combustion, operates at high primary zone temperatures, circa 2,500 K, resulting in high thermal NO_x formation. Lowering the flame temperature, and hence NO_x production, can be achieved by injection of **diluents** such as water or steam into the primary zone, which quench the flame and has been

① 低发热量，指燃料中的水分在燃烧过程结束后以水蒸气形式存在时的热量。

② 华白数，又称华白指数，指在规定参比条件下的体积高位发热量除以在相同的规定计量参比条件下的相对密度的平方根。

successfully employed for many years by many of the gas turbine manufacturers. Different OEMs use differing methods for water or steam injection, but all recognize the impact each has on reliability and life cycle costs. Generally, such combustion systems have been more tolerant to different fuel types.

10. Reducing exhaust emissions by injecting water or steam into the primary zone is compared to the benefits of DLE/DLN solution. Wet injection needs a large quantity of **demineralized** water and there is an impact on the service **regime**, with more frequent planned interventions. The ratio of water, or steam, to fuel used results in lower NO_x, but can impact CO emissions in a **detrimental** manner. Completion of a retrofit of existing gas turbines was made to include DLE combustion achieving a notable environmental benefit, offering almost a 5 times reduction in NO_x emissions compared to previous abatement system employed.

(Source: http://www.power-eng.com/articles/print/volume-118/issue-5/features/fuels-combustion-environmental-considerations-in-industrial-gas-turbines.html)

New Words and Expressions

abatement	*n.*	控制，减轻
contaminant	*n.*	污染物
detrimental	*adj.*	不利的；有害的
demineralized	*adj.*	去矿物质的
dew	*n.*	露水
diluent	*n.*	稀释气体，稀释剂
inert	*adj.*	惰性的
monoxide	*n.*	一氧化物
nitrogen	*n.*	氮
norm	*n.*	标准，规范
regime	*n.*	管理体制
stewardship	*n.*	管理工作
unburnt	*adj.*	未燃的

Comprehension Checks

Work in pairs and discuss the following questions. Then share your ideas with the rest of the class.

1. What method is applied to gas turbines for minimizing possible negative effect on environment? (para.1)

2. Could you state some of the factors that might guarantee the smooth operation in a modern gas turbine? (para.2)
3. What is the added value of fuels in GT application? (para.4)
4. Could you briefly introduce different types of combustion emissions mentioned in the passage? (para.6-8)
5. What method is widely employed to reduce the temperature of NO_x production? (para.9)
6. What benefit would the retrofit of existing gas turbines bring along? (para.10)

Vocabulary Tasks

I. Complete the sentences below. The first letter of the missing word has been given.

1. Moisture in the atmosphere condensed into d________ during the night.
2. Copper is a c________ that can mess up nonmetallic stages of the manufacturing process, so the machines that add copper need to be carefully segregated.
3. Papain of a higher digestive power may be reduced to the official standard by admixture with papain of lower activity, lactose, or other suitable d________.
4. Generally, a two-stroke engine produces a lot of power for its size because there are twice as many c________ cycles occurring per rotation.
5. Under the new r________, all sheep and cattle will be regularly tested for disease.

II. Fill in the blanks with the words given below. Change the forms where necessary.

detrimental	retrofit	abatement	norm	demineralized
inert	diluent	combustion	contaminant	parameter

1. Economists consider deflation to be ________ to the economy.
2. Consequently, low emissions became the ______ and not the exception, especially for pollutants such as nitrogen oxide (NO_x), carbon monoxide (CO) and unburnt hydrocarbons (UHC).
3. Completion of a ________ of existing gas turbines was made to include DLE combustion achieving a notable environmental benefit, offering almost a 5 times reduction in NO_x emissions compared to previous abatement system employed.
4. Wet injection needs a large quantity of ________ water and there is an impact on the service regime, with more frequent planned interventions.
5. Forests and plants remove carbon from the atmosphere, potentially accounting for more than 40% of carbon ________ opportunities between now and 2020.
6. The severity of the risk depends on the radionuclide mix and the level of ________ released.

7. Foodstuffs may be dried in air, superheated steam, in vacuum, in ________ gas, and by the direct application of heat.
8. A linear correlation is found between the ratio of unstretched laminar burning velocity with and without the ________ and the dilution ratio.
9. Climate sensitivity is a key ________ for estimating the long-term temperature increase expected from a doubling of the concentration of greenhouse gas emissions in the atmosphere.
10. A major disadvantage of coal ________ has been air pollution by soot, dust, ash, exhaust and large amounts of the greenhouse gas carbon dioxide.

ext B

Reverse Combustion: Can CO_2 Be Turned Back into Fuel?

1. In the 1990s a graduate student named Lin Chao at Princeton University decided to bubble carbon dioxide into an electrochemical cell. Using cathodes made from the element **palladium** and a catalyst known as **pyridinium** — a garden variety organic chemical that is a by-product of oil refining — he discovered that applying an electric current would assemble methanol from the CO_2. He published his findings in 1994 — and no one cared.

2. But by 2003, Chao's successor in the Princeton lab of chemist Andrew Bocarsly was deeply interested in finding a solution to the growing problem of the CO_2 pollution causing global climate change. Graduate student Emily Barton picked up where he left off and, using an electrochemical cell that employs a **semiconducting** material used in **photovoltaic** solar cells for one of its **electrodes**, succeeded in tapping sunlight to transform CO_2 into the basic fuel.

3. Turning CO_2 into fuels is exactly what photosynthetic organisms have been doing for billions of years, although their fuels tend to be foods, like sugars. Now humans are trying to store the energy in sunlight by making a liquid fuel from CO_2 and hydrogen — a prospect that could recycle CO_2 emissions and slow down the rapid buildup of such greenhouse gases in the atmosphere. "You take electricity and combine CO_2 with hydrogen to make gasoline," explained Arun Majumdar, director of the Advanced Research Projects Agency-Energy (ARPA-E)① that is pursuing such technology, at a conference in March. "This is like killing four birds with one stone" — namely, energy security, climate change, the federal deficit and, potentially, unemployment.

4. In fact, the problem with turning CO_2 back into a hydrocarbon fuel is not so much in

---

① 高级能源研究计划署。

transformation — there are at least three potential approaches to do so with sunlight, along with a process that employs high pressures and temperatures, so-called Fischer-Tropsch reaction[①], which is currently used — but rather the tremendous expense involved. "It's an uphill process to convert CO_2 to methanol. It's going to cost you some energy," Bocarsly says. "The current rate of generating methanol is not high enough to be commercial."

Solar Heat

5. The sun bathes Earth in more energy in an hour than human civilization uses in a year. Giant dish mirrors in the New Mexico desert **erected** by scientists at Sandia National Laboratories[②] capture some of that energy and have been used to concentrate it on a cylindrical machine that looks like a beer **keg** — a would-be solar-fuel generator. Nestled inside that machine are a series of rotating, concentric rings. Turning at roughly one rotation per minute, the CR_5 — for counter rotating ring receiver reactor recuperator — moves these rings **studded** with teeth containing iron oxide (also known as **ferrite**, or **rust**) or **cerium** oxide (ceria) into and out of the sunlight. The sun heats the teeth as high as 1,500 ℃ — driving oxygen out of the rust — before they rotate back into the darkness and cool off to roughly 900 ℃. In the darkness, steam or CO_2 is injected and the greedy ferrite sucks the oxygen out of those molecules — leaving carbon monoxide (CO) or hydrogen (H_2) behind — before rotating back into the sun.

6. The resulting $CO\text{–}H_2$ mixture is so-called synthesis gas, or syngas — the basic molecular building block of fossil fuels, chemicals, even plastics. The CR_5 "is a chemical heat engine", says chemical physicist Ellen Stechel, program manager of Sandia's Sunshine to Petrol project that basically seeks to reverse fossil-fuel combustion. "It's doing chemical work, breaking a bond."

7. The CR_5 has already been turned on three times — and will likely be run again this year before the sun gets too weak, according to chemical engineer James Miller of Sandia, co-inventor of the device. But it has never reached the steady state necessary to efficiently throw off syngas. The problem is the thousands of ceramic tiles that form the reactive teeth on the edge of the rotating rings, some of which break as the process heats up. "You're cycling back and forth from 1,500 to 900 ℃, and that's a lot to ask of a material," notes chemist Gary Dirks, director of LightWorks at Arizona State University, who is not involved with the project.

---

① 费托反应是以合成气（CO 和 H_2）为原料在催化剂（主要是铁系）和适当反应条件下合成以石蜡烃为主的液体燃料的工艺过程。

② 桑迪亚国家实验室。

Artificial Photosynthesis

8. Nature has an answer. Plants pull CO_2 out of the air and, thanks to the ongoing burning of the results of millions of years of photosynthesis (otherwise known as fossil fuels), atmospheric concentrations continue to rise. Unfortunately, plants are **woefully** inefficient at turning sunlight into food — averaging at best 1 percent of the incoming sunlight stored as chemical energy, thanks to competing concerns like survival — one main reason that the U.S. Department of Energy (DoE) estimates that at best, 15 percent of the nation's energy needs could come from biofuels.

9. Chemist Nathan Lewis of the California Institute of Technology would like to improve on that by mimicking the processes of photosynthesis — light absorbers, molecule-makers and **membranes** to separate various products, among other things — artificially. "Nature uses **enzymes**; we use inorganic complexes or metals," Lewis says of a new effort to create artificial photosynthesis launched with DoE funding on July 22. "Materials carry as much electric current as you like because they move electrons rather than molecules."

(Source: https://www.scientificamerican.com/article/turning-carbon-dioxide-back-into-fuel/)

New Words and Expressions

cerium	*n.*	铈
electrode	*n.*	电极
erect	*v.*	使竖立，建造
enzyme	*n.*	酶
ferrite	*n.*	铁素体，铁氧体
keg	*n.*	小桶
membrane	*n.*	膜
palladium	*n.*	钯
photovoltaic	*adj.*	光电的
pyridinium	*n.*	吡啶盐
rust	*n.*	锈
semiconducting	*adj.*	半导体的
stud	*v.*	镶嵌
woefully	*adv.*	悲伤地

Comprehension Questions

1. Which of the following is NOT essential for converting CO_2 into fuel?
   A. Cathodes. B. Pyridinium. C. Methanol.
2. According to the passage, what's the major problem in turning CO_2 back into fuel?
   A. The converting process involves high pressures and temperature.
   B. The current approach for generating fuel is not commercial, because it is costly.
   C. Converting CO_2 to methanol would cost some energy.
3. Which factor is mainly responsible for the failure in steadily generating syngas?
   A. Temperature. B. Pressure. C. CR_5.
4. Which of the following is TRUE about the photosynthesis?
   A. Plants are able to converting solar energy into food at a very high rate.
   B. It makes the DoE believe that more than 15 percent of the nation's energy could be provided by biofuels.
   C. Scientists are paying efforts to simulating the process of photosynthesis in an advanced approach.
5. What is the author's overall attitude toward the future of reverse combustion?
   A. Positive. B. Negative. C. Impersonal.

Grammar Coach

Run-Ons, Fragments and Comma Splice

连写句、句子片段与“一逗到底”是英语写作中常见的三种错误。这三种错误在科技英语文本中出现，会影响读者对文本内容的理解。因此，识别并学会改正这三种错误对于英语学习十分重要。下面将对这三种错误的概念及每种错误的修改方法进行介绍。

1. 连写句

连写句（run-ons）指的是错误地将两个或两个以上的独立分句写在一个句子里，没有使用正确的标点符号或连词分割句意的语法现象。在修改此类错误时，先要分割句子，把连写句分成几个独立的主句；然后，采用分号或者句号分割不同的主句。如果句子之间有逻辑联系，可添加适当的连词连接各个主句。

例如： Some of the customers in the restaurant like hot oatmeal others prefer a packaged cold cereal.

错误诊断：run-ons

错误修改：用分号、句号分割句子或添加适当的连词连接两个主句，修改如下。

a. Some of the customers in the restaurant like hot oatmeal; others prefer a packaged cold cereal.

b. Some of the customers in the restaurant like hot oatmeal. Others prefer a packaged cold cereal.

c. Some of the customers in the restaurant like hot oatmeal, while others prefer a packaged cold cereal.

2. 句子片段

句子片段（fragments）指的是不完整的句子、有残缺的句子，即形式上像一个完整的句子，但句意不完整或语法成分不完整。语言学家认为一个句子通常应该具有完整的句意及完整的语法成分等。在修改此类错误时，第一要判断好句子的语法成分是否完整；第二要判断句意是否完整，理清句子之间的逻辑关系，添加恰当的标点符号和连接词，从而达到句意完整且符合逻辑关系。

例如：

（1）Whenever the traffic gets heavy.

错误诊断：fragments（语法成分不完整：句子中只提供了时间状语，缺少主句；语意不完整）

错误修改：根据语意，添加主句，修改如下。

Whenever the traffic gets heavy, measures should be taken to deal with the situation.

（2）Find a parking space there is usually easy during the week.

错误诊断：fragments（语法成分错误，句中有两个谓语动词：find 和 is，句式杂糅，缺少主语）

错误修改：将 find 改为 finding，即动名词作主语，修改如下。

Finding a parking space there is usually easy during the week.

3. "一逗到底"

"一逗到底"（comma splice）指的是只用逗号连接两个或两个以上完整的句子，句子之间没有任何的连接词的语法现象。这是一种常见的语法错误。在修改此类错误时，可以采用如下三种方法。

例如：Water resources are scarce, we should value them, we should also develop other resources. （×）

错误诊断：comma splice

错误修改：用句号将句子之间隔开，分隔为互相独立的句子；用分号将句与句之间隔开，分隔为互相独立的句子；在句子之间加上连接词，如 and、but、or、because、so、yet 等，这样不仅符合语法规则，也交代了句子之间的逻辑关系，修改如下。

a. Water resources are scarce. We should value them. We should also develop other resources.

b. Water resources are scarce; we should value them; we should also develop other resources.

c. Water resources are scarce so we should value them, and we should develop other resources.

Exercise

Please identify and correct possible errors in the following sentences.

1. The government striving to achieve political, economic, social and educational equality.
2. Some of us work hard to learn vocabulary and grammar rules others pay no efforts to English study.
3. A steady breeze is blowing today, kite fliers are gathering in the park.
4. Books that are about skating and other winter sports.
5. Most professional basketball players are tall and rangy, Squirt Simpson, is a notable exception.
6. The border patrol inspected every vehicle for non-registered visitors traffic was backed up for hours.
7. The plane left early in the morning, he did not wake up in time.
8. She was driving along the road looking for a place to park.

References

［1］STEVE H. Written English: a guide for electrical and electronic students and engineers[M]. Boca Raton: CRC Press, 2015.

［2］JESPERSEN O. Essentials of English grammar［M］. London: Routledge, 2006.

［3］车德勇，孙佰仲，魏高升，等. 能源动力类专业英语［M］. 北京：中国电力出版社，2011.

［4］陈忠华. 科技英语教学的理论与实践［M］. 石家庄：河北科学技术出版社，1990.

［5］何岑成，承红. 科技英语写作指导教程［M］. 北京：北京工业大学出版社，2012.

［6］江涛. 能源行业英语［M］. 北京：中国海关出版社, 2008.

［7］卜玉坤，金晶鑫，鲁晖，等. 能源动力英语：2［M］. 北京：外语教学与研究出版社, 2013.

［8］秦荻辉. 科技英语写作高级教程［M］. 2 版. 西安：西安电子科技大学出版社，2011.

［9］秦荻辉. 科技英语语法［M］. 北京：外语教学与研究出版社，2007.

［10］秦荻辉. 实用科技英语写作技巧［M］. 上海：上海外语教育出版社，2001.

［11］田瑞，闫素英. 能源与动力工程概论［M］. 北京：中国电力出版社，2008.

［12］谢诞梅，陈启卷，金振齐. 能源动力类专业英语［M］. 北京：中国水利水电出版社，2009.

［13］斯韦尔斯. 科技英语写作技巧［M］. 西安：陕西人民出版社,1985.

［14］余樟亚，庄起敏. 电力新能源英语［M］. 北京：国防工业出版社，2009.

［15］王建武，李民权，曾小珊. 科技英语写作：理论、技巧、范例［M］. 西安：西北工业大学出版社，2003.

［16］章振邦. 新编英语语法教程：学生用书［M］. 6 版. 上海：上海外语教育出版社，2017.

［17］张链，汪磊. 新能源新材料专业英语基础教程［M］. 北京：中国科学技术大学出版社，2012.

［18］赵俊英. 现代英语语法大全［M］. 北京：商务印书馆，2016.